Proof and the Art of Mathematics:
Examples and Extensions

Proof and the Art of Mathematics:
Examples and Extensions

Joel David Hamkins

The MIT Press
Cambridge, Massachusetts
London, England

The MIT Press would like to thank the anonymous peer reviewers who provided comments on drafts of this book. The generous work of academic experts is essential for establishing the authority and quality of our publications. We acknowledge with gratitude the contributions of these otherwise uncredited readers.

This book was set using LaTeX and TikZ by the author. Printed and bound in the United States of America.

ISBN: 978-0-262-54220-3

10 9 8 7 6 5 4 3 2

Contents

Contents

Preface

The best way to learn mathematics is to dive in and do it. Don't just listen passively to a lecture or read a book—you have got to take hold of the mathematical ideas yourself! Mount your own mathematical analysis. Formulate your own mathematical assertions. Consider your own mathematical examples. I recommend play—adopt an attitude of playful curiosity about mathematical ideas; grasp new concepts by exploring them in particular cases; try them out; understand how the mathematical constructions from your proofs manifest in your examples; explore all facets, going beyond whatever had been expected. You will find vast new lands of imagination. Let one example generalize to a whole class of examples; have favorite examples. Ask questions about the examples or about the mathematical idea you are investigating. Formulate conjectures and test them with your examples. Try to prove the conjectures—when you succeed, you will have proved a theorem. The essential mathematical activity is to make clear claims and provide sound reasons for them. Express your mathematical ideas to others, and practice the skill of stating matters well, succinctly, with accuracy and precision. Don't be satisfied with your initial account, even when it is sound, but seek to improve it. Find alternative arguments, even when you already have a solid proof. In this way, you will come to a deeper understanding. Test the statements of others; ask for further explanation. Look into the corner cases of your results to probe the veracity of your claims. Set yourself the challenge either to prove or to refute a given statement. Aim to produce clear and correct mathematical arguments that logically establish their conclusions, with whatever insight and elegance you can muster.

This book is offered as a companion volume to my book *Proof and the Art of Mathematics*, which I have described as a mathematical coming-of-age book for students learning how to write mathematical proofs. Spanning diverse topics from number theory and graph theory to game theory and real analysis, *Proof and the Art* shows how to prove a mathematical theorem, with advice and tips for sound mathematical habits and practice, as well as occasional reflective philosophical discussions about what it means to undertake mathematical proof. In *Proof and the Art*, I offer a few hundred mathematical exercises, challenges to the reader to prove a given mathematical statement, each a small puzzle to figure out; the intention is for students to develop their mathematical skills with these challenges of mathematical reasoning and proof.

Here in this companion volume, I provide fully worked-out solutions to all of the odd-numbered exercises, as well as a few of the even-numbered exercises. In many cases, the solutions here explore beyond the exercise question itself to natural extensions of the ideas. My attitude is that, once you have solved a problem, why not push the ideas harder to see what further you can prove with them? These solutions are examples of how one might write a mathematical proof. I hope that you will learn from them; let us go through them together. The mathematical development of this text follows the main book, with the same chapter topics in the same order, and all theorem and exercise numbers in this text refer to the corresponding statements of the main text. This book was typeset using LaTeX, and all figures were created using TikZ in LaTeX, except in chapter 12 for the Königsberg bridge image, which I drew myself by hand, and the triangulated torus image, released by user AG2gaeh under a Creative Commons license.

Joel David Hamkins
January 2020

About the Author

I am an active research mathematician and mathematical philosopher at Oxford University. I work on diverse topics in mathematical logic and the philosophy of mathematics, including especially the mathematics and philosophy of the infinite. For me, mathematics is a lifelong process of learning and exploring. Truly one of life's great joys is to share interesting new mathematical ideas or puzzles with others, and I find myself doing so not only in my research papers and books but also on my blog, on Twitter, and on MathOverflow. I cordially invite you to join the conversations on all these forums (links below). My new book, *Lectures on the Philosophy of Mathematics*, forthcoming with MIT Press, emphasizes a mathematically grounded perspective on the philosophy of mathematics, an approach that I believe will appeal both to mathematicians and to philosophers of mathematics.

Joel David Hamkins
Professor of Logic *&* Sir Peter Strawson Fellow in Philosophy
Oxford University, University College
High Street, Oxford OX1 4BH

joeldavid.hamkins@philosophy.ox.ac.uk
joeldavid.hamkins@maths.ox.ac.uk
Blog: http://jdh.hamkins.org
MathOverflow: http://mathoverflow.net/users/1946/joel-david-hamkins
Twitter: @JDHamkins

1 A Classical Beginning

The classical theorem that $\sqrt{2}$ is irrational is a gem of antiquity, proved by a beautiful argument that has endured millennia, a pinnacle of human insight and achievement. Here, we begin with exercises that solidify the foundations, establishing facts used in the proof of this classic result, and then move on to exercises that generalize the result beyond $\sqrt{2}$, proceeding in small steps that achieve the same conclusion in more and more cases, until ultimately we provide a complete general criterion for when $\sqrt{n}$ is irrational.

1.1 Prove that the square of any odd number is odd.

Following the theorem-proof format, let us make the claim here as a formal theorem statement, for which we then provide proof.

Theorem. *If n is an odd integer, then n^2 also is odd.*

Proof. Assume that n is an odd integer. By definition, this means that $n = 2k + 1$ for some integer k. In this case, we can calculate that

$$n^2 = (2k + 1)^2 = (2k + 1)(2k + 1) = 4k^2 + 4k + 1.$$

We can write this final sum as $2(2k^2 + 2k) + 1$, which is $2r + 1$, if we let $r = 2k^2 + 2k$. And so n^2 also is odd, as desired. □

More generally, this theorem is a consequence of the following familiar fact:

Theorem. *The product of any two odd integers is odd.*

Proof. Suppose that n and m are odd integers. By definition, this means that $n = 2k + 1$ and $m = 2r + 1$ for some integers k and r. Now observe that

$$nm = (2k + 1)(2r + 1) = 4kr + 2k + 2r + 1.$$

This is equal to $2(2kr + k + r) + 1$, which is 2 times an integer, plus 1; and so nm is odd, as desired. □

> **1.3** Prove that $\sqrt[4]{2}$ is irrational. Give a direct argument, but kindly also deduce it as a corollary of theorem 1 in the main text.

Theorem. $\sqrt[4]{2}$ *is irrational.*

Proof as direct argument. Suppose toward contradiction that $\sqrt[4]{2}$ is rational. In this case, we may express it as a fraction

$$\sqrt[4]{2} = \frac{p}{q},$$

where p and q are integers, and q is not zero. We may furthermore assume that this fraction is in lowest terms, so that p and q have no common factors. By raising both sides to the 4th power, we conclude that

$$2 = \frac{p^4}{q^4}$$

and therefore that $2q^4 = p^4$. It follows that p^4 is even. Since the product of any number of odd numbers remains odd, it follows that p cannot be odd, and so p is even. So $p = 2k$ for some integer k, and consequently, $2q^4 = p^4 = (2k)^4 = 16k^4$, which implies that $q^4 = 8k^4$. So q^4 also is even, and so q is even. So the fraction p/q was not in lowest terms after all, contradicting our assumption. So $\sqrt[4]{2}$ cannot have been rational in the first place, and so it is irrational. □

Alternative proof as corollary to theorem 1. Theorem 1 in the main text was the assertion that $\sqrt{2}$ is irrational. The key thing to notice is that $\sqrt{2}$ is simply the square of $\sqrt[4]{2}$,

$$\sqrt{2} = \left(\sqrt[4]{2}\right)^2,$$

and so if $\sqrt[4]{2}$ were a rational number p/q, then $\sqrt{2}$ would be the square of this number $\sqrt{2} = (p/q)^2 = p^2/q^2$, which remains rational. This would contradict theorem 1. So we conclude that $\sqrt[4]{2}$ cannot be rational. □

We can generalize this observation to the following theorem:

Theorem. *If r is irrational, then so is $\sqrt{r}$ and indeed $\sqrt[k]{r}$ for any positive integer k.*

The exercise is a special case of this, since $\sqrt[4]{2} = \sqrt{\sqrt{2}}$.

Proof. Notice that $r = \left(\sqrt[k]{r}\right)^k$, and so if $\sqrt[k]{r}$ were a rational number p/q, then we would have $r = (p/q)^k = p^k/q^k$, which is rational, contrary to our assumption. □

1.5 Prove that $\sqrt{5}$ and $\sqrt{7}$ are irrational. Prove that $\sqrt{p}$ is irrational, whenever p is prime.

Let's warm up with a direct argument for $\sqrt{5}$ and then generalize it to $\sqrt{p}$ for arbitrary primes. We shall make use of the fundamental theorem of arithmetic, stating that every positive integer has a unique prime factorization.

Theorem. $\sqrt{5}$ *is irrational.*

Proof. Suppose toward contradiction that $\sqrt{5} = n/m$ is rational, represented by a fraction with integers n and $m \neq 0$. By squaring both sides and clearing the denominator, we see that

$$5m^2 = n^2.$$

From this, it follows that n^2 is a multiple of 5. This implies that n itself must be a multiple of 5, since the only way to get 5 into the prime factorization of n^2 is to have it already in n, as the prime factorization of n^2 is obtained from the prime factorization of n by squaring every term. And so $n = 5k$ and thus $5m^2 = n^2 = (5k)^2 = 25k^2$. So we can cancel 5 and deduce $m^2 = 5k^2$. So 5 must also appear in the prime factorization of m^2 and hence also in that of m. So n/m was not in lowest terms after all, contrary to our assumption. So $\sqrt{5}$ must be irrational. $\square$

Essentially the same argument works with any prime number, since the only thing we had used about the number 5 in the argument was that it was prime. So let us set out and prove this more general fact.

Theorem. *If p is a prime number, then $\sqrt{p}$ is irrational.*

Proof. Let p be a prime number, and suppose toward contradiction that $\sqrt{p}$ is rational. So we may represent it as a fraction

$$\sqrt{p} = \frac{n}{m},$$

where n and m are integers, and $m \neq 0$. By squaring both sides, we see that $p = n^2/m^2$ and consequently $pm^2 = n^2$. It follows that p appears in the prime factorization of n^2. Therefore, it must also appear in the prime factorization of n. So $n = pk$, and so $pm^2 = n^2 = (pk)^2 = p^2k^2$. By canceling one p, we deduce that $m^2 = pk^2$. And so p must appear in the prime factorization of m^2, and consequently also in that of m. So both n and m are multiples of p, which contradicts our assumption that p/q was in lowest terms. So $\sqrt{p}$ must be irrational. $\square$

We may now deduce the original cases of the question as a corollary.

Corollary. $\sqrt{5}$ *and* $\sqrt{7}$ *are each irrational.*

Proof. This is an instance of the theorem, since 5 and 7 are each prime.   □

One might object that we needn't have proved the first theorem above, that $\sqrt{5}$ is irrational, since we've just now deduced it as a corollary to the more general theorem that $\sqrt{p}$ is irrational for every prime p. But I would find that objection mistaken. Just because you can deduce a theorem as a consequence of a more general theorem doesn't mean that you should only do the argument that way. The earlier, more elementary proof retains value simply because it is a more elementary or concrete instance, one that furthermore exhibits the main idea that led to the more general theorem in the first place. It was easier to understand the first argument simply because it has used the particular number 5 instead of the variable p, which meant one fewer abstraction woven into the argument. In addition, let me say categorically that there is absolutely nothing wrong with proving the same fact twice or more times with different arguments. I find it to be good mathematical style, even in a formal mathematical research paper, to warm up with an easier case of what will ultimately be a more general or abstract argument.

> **1.7**   Prove that $\sqrt{2m}$ is irrational, whenever m is odd.

For example, $\sqrt{10}$, $\sqrt{14}$, $\sqrt{30}$, and $\sqrt{50}$ are each irrational.

Theorem. $\sqrt{2m}$ *is irrational, whenever m is odd.*

Proof. Suppose toward contradiction that m is odd and that $\sqrt{2m} = p/q$ is rational, where p and $q \neq 0$ are integers, and where this fraction is in lowest terms. Squaring both sides leads to $2mq^2 = p^2$. Therefore, p^2 is even, and so p also must be even. So $p = 2k$ and thus $2mq^2 = p^2 = (2k)^2 = 4k^2$. Canceling the 2 leads to $mq^2 = 2k^2$. So mq^2 is even. But since m is odd, this means that q^2 must be even, and so q also is even. So p/q was not in lowest terms after all, and so $\sqrt{2m}$ cannot have been rational.   □

> **1.9**   Criticize this "proof." Claim. $\sqrt{n}$ is irrational for every natural number n.
> Proof. Suppose toward contradiction that $\sqrt{n} = p/q$ in lowest terms. Square
> both sides to conclude that $nq^2 = p^2$. So p^2 is a multiple of n, and therefore p
> is a multiple of n. So $p = nk$ for some k. So $nq^2 = (nk)^2 = n^2k^2$, and therefore
> $q^2 = nk^2$. So q^2 is a multiple of n, and therefore q is a multiple of n, contrary to
> the assumption that p/q is in lowest terms. □

Let us criticize the proposed proof, which is not correct. To begin, let us consider carefully the claim that is being made.

"Claim." $\sqrt{n}$ *is irrational for every natural number n.*

But this claim is not true! We can easily find many counterexamples. Consider the case $n = 25$, for example, for which $\sqrt{n} = \sqrt{25} = 5$, which is certainly a rational number, and similarly $\sqrt{49} = 7$ and $\sqrt{100} = 10$. So sometimes $\sqrt{n}$ is rational, and these are counterexamples that refute the claim. It follows, of course, that the proof cannot be right, and so we should expect to find some kind of mistake. But where exactly does the argument go wrong? Let's go through each sentence of the argument.

"Proof." *Suppose toward contradiction that* $\sqrt{n} = p/q$ *in lowest terms.*

The proof starts out completely fine. It seems that we shall try to prove the claim by contradiction, by supposing that $\sqrt{n}$ is rational and then trying to derive a contradiction from this assumption.

Square both sides to conclude that $nq^2 = p^2$.

This step also is fine; it follows similar reasoning as we used when proving the theorem on $\sqrt{2}$ and the other cases.

So p^2 *is a multiple of n, and therefore p is a multiple of n.*

Yes, p^2 is a multiple of n, since we observed above that $nq^2 = p^2$, so the first part of this is correct. But the next statement, "and therefore p is a multiple of n" is not right. Just because a number p^2 is a multiple of a number, it doesn't follow that p must be a multiple of the number. For example, 100 is a multiple of 25, but 10 is not. In the main argument about $\sqrt{2}$, we did argue that if p^2 is a multiple of 2, then p also is a multiple of 2. And it is fine in that case, but what makes it correct is that 2 is a prime number, and so if it shows up in the prime factorization of p^2, then it must have been already in the factorization of p itself. In general, if p^2 is a multiple of a number n, this doesn't mean that p has to be a multiple of n. So this is where the argument is incorrect.

So $p = nk$ *for some k. So* $nq^2 = (nk)^2 = n^2k^2$, *and therefore* $q^2 = nk^2$.

If we did know that p was a multiple of n, then this part would be correct.

So q^2 *is a multiple of n, and therefore q is a multiple of n, contrary to the assumption that p/q is in lowest terms.* □

This is the same mistake as earlier, but with q instead of p. We can't conclude in either case that p or q is a multiple of n, and so no contradiction is reached after all; the proof is not valid.

1.11 For which natural numbers n is $\sqrt{n}$ irrational? Prove your answer.

Our answer to this question will generalize and unify all the answers we gave above, which will become special cases of this general result. This process therefore illustrates a frequent pattern in mathematics, where one finds the general argument only after having made arguments in various special cases. Faced with a mathematical question or collection of questions, one proves at first what one can about it; and when the argument is therefore grasped more fully as a result of this progress, one often realizes that a slightly more general argument can prove a more general fact; and then a still more general argument proves a still more general fact, generalizing again and again. Eventually, in the fortunate cases, one arrives at a satisfying summative result, which unifies the earlier arguments while using essentially similar reasoning, reaching the unifying general fact to which the argument was tending, from which all the earlier results follow as immediate special cases. That is the pattern of this case, where at first we proved $\sqrt{2}$ is irrational, but then generalized to $\sqrt{p}$ for prime numbers p, and $\sqrt{2m}$ where m is odd. What is it, really, that is making the arguments work? We can generalize to the following general account.

Theorem. *For any natural number n, the number $\sqrt{n}$ is rational if and only if n is a perfect square. That is, $\sqrt{n}$ is rational if and only if $n = k^2$ for some natural number k, in which case $\sqrt{n} = k$.*

In particular, it follows that $\sqrt{n}$ is a rational number if and only if it is itself a natural number. Much of the argument is contained in the following lemma.

Lemma. *A natural number n is a perfect square if and only if every exponent in the prime factorization of n is even.*

Proof. For the forward direction, notice that if one has a perfect square $n = k^2$, then the prime factorization of n is obtained by simply squaring the prime factorization of k, and this will result in all even exponents. That is, if we have the prime factorization of k as

$$k = p_1^{e_1} p_2^{e_2} \cdots p_r^{e_r},$$

then the prime factorization of the square is

$$n = k^2 = \left(p_1^{e_1} p_2^{e_2} \cdots p_r^{e_r} \right)^2 = p_1^{2e_1} p_2^{2e_2} \cdots p_r^{2e_r},$$

and this has all exponents even.

Conversely, if the prime factorization of a natural number n has all even exponents

$$n = p_1^{2e_1} p_2^{2e_2} \cdots p_r^{2e_r},$$

then by the same algebraic manipulation we used above, it follows that $n = k^2$, where

$$k = p_1^{e_1} p_2^{e_2} \cdots p_r^{e_r}.$$

And so n is a perfect square if and only if all the exponents in the prime factorization of n are even. □

We can now prove the theorem.

Proof of theorem. The converse direction is immediate, since if $n = k^2$ for natural numbers n and k, then $\sqrt{n} = k$, which is rational. The real content of the theorem is that this trivial case is the only way that $\sqrt{n}$ can be rational.

For the forward direction, suppose that $\sqrt{n}$ is rational. So we may express it as a fraction $\sqrt{n} = r/s$, where r and s are integers and $s \neq 0$. By squaring both sides and clearing the denominator, we see that $ns^2 = r^2$. Since r^2 is a perfect square, all the exponents in the prime factorization of r^2 will be even. And so all the exponents in the prime factorization of ns^2 are even. But since the exponents coming from s^2 will be even, it follows that the remaining exponents, the exponents in the prime factorization of n, must also be even, in order to combine to make all even exponents in ns^2. And so by the lemma, n must be a perfect square. □

1.13 Prove that the irrational real numbers are exactly those real numbers that are a different distance from every rational number. Is it also true if you swap "rational" and "irrational"?

Theorem. *A real number is irrational if and only if it has a different distance to every rational number.*

Proof. ($\rightarrow$) We prove the contrapositive. Suppose that x has the same distance to two distinct rational numbers r and s. So x must be exactly halfway between r and s, and so it is the average $x = (r + s)/2$; thus, it is rational.

($\leftarrow$) Again, we prove the contrapositive. If x is rational, then so are $x + 1$ and $x - 1$, and x differs from each of them by 1. So x is the same distance from two distinct rational numbers. □

Meanwhile, the statement is not true if you swap "rational" and "irrational," since the number 0 is rational, but it has the same distance to π as to $-\pi$.

More generally:

Theorem. *Every real number is equidistant from two irrational numbers.*

Proof. If x is rational, then consider $x + \sqrt{2}$ and $x - \sqrt{2}$, both of which are irrational. If x is irrational, then consider $x + 1$ and $x - 1$, both of which are irrational. So in any case, x is equidistant from two irrational numbers. $\square$

2 Multiple Proofs

Mathematical progress is often achieved when one explores alternative proofs for a theorem. A different argument may reveal different aspects of the problem or different avenues of generalization. Here, we explore several such aspects arising from the various alternative proofs given in the main text of the simple fact that $n^2 - n$ is always even for any natural number.

> **2.1** Prove that the sum, difference, and product of two even numbers is even. Similarly, prove that the sum and difference of two odd numbers is even, but the product of odd numbers is odd.

To provide a solution to this exercise, we need first to know exactly what it means to be an even number or an odd number. What is the definition? The *even* integers are those that are a multiple of 2, the numbers of the form $2k$, where k is an integer. The *odd* integers have the form $2k + 1$, with a remainder of 1 when dividing by 2. By the Euclidean algorithm, every number has a remainder either of 0 or 1 when dividing by 2, and so every number is either even or odd, but never both.

Theorem. *The sum, difference, and product of even numbers remain even.*

Proof. This is quite easy to see. Suppose we have two even numbers, say, $2k$ and $2r$. If we add them, we get $2k + 2r = 2(k + r)$, which is even because it is a multiple of 2. If we subtract, we get $2k - 2r = 2(k - r)$, which is even because it is a multiple of 2. And if we multiply them, we get $2k \cdot 2r = 2(2kr)$, which is even because it is a multiple of 2. So the sum, difference, and product of any two even numbers remain even. $\square$

Theorem. *The sum and difference of two odd numbers are even, but the product of two odd numbers is odd.*

Proof. Suppose that we have two odd numbers, $2k + 1$ and $2r + 1$. If we add them, we get

$$(2k + 1) + (2r + 1) = 2k + 2r + 2 = 2(k + r + 1),$$

which is even, because it is a multiple of 2. If we subtract them, we get

$$(2k + 1) - (2r + 1) = 2k - 2r = 2(k - r),$$

which is even, because it is a multiple of 2. But if we multiply them, we get

$$(2k + 1)(2r + 1) = 2k \cdot 2r + 2k + 2r + 1 = 2(2kr + k + r) + 1,$$

which is odd because it is a multiple of 2 plus 1.   □

This latter argument was the same we had considered in exercise 1.1 as a generalization of the claim that the square of any odd number is odd.

2.3 True or false: if the product of one pair of positive integers is larger than the product of another pair, then the sum also is larger.

The question is whether $pq > rs$ implies $p + q > r + s$ in the positive integers.

Theorem. *The statement is false. It is not always true that if the product of one pair of positive integers is larger than the product of another pair, then the sum also is larger.*

Proof. To prove that the statement is false, it suffices to give a counterexample. Consider the two pairs $2, 3$ and $1, 5$. The product of the first pair is larger than the product of the second pair, because $2 \cdot 3 = 6 > 5 = 1 \cdot 5$, but the sum of the first pair is only $2 + 3 = 5$, whereas the sum of the second pair is $1 + 5 = 6$. So this pair of pairs is a counterexample to the statement, and therefore the statement is not true in general.   □

To refute a universal statement, it suffices to give a particular counterexample, and for clarity it is often best to give a very specific counterexample when possible.

2.5 Prove that the product of k consecutive integers is always a multiple of k.

Theorem. *The product of k consecutive integers is always a multiple of k.*

This theorem exhibits a common situation in mathematics, where one proves a theorem by one argument, perhaps by a comparatively straightforward argument, but actually, a considerably stronger result can be proved, by a somewhat trickier argument. In this case, we have the simple result here and a much stronger result, which I shall explain after exercise 2.7. For now, let's prove just the result that is claimed here.

Proof. Since every kth integer is a multiple of k, it follows that in any block of k consecutive integers exactly one of them must be a multiple of k. Therefore, the product must also be a multiple of k. ☐

2.7 Prove that the product of any four consecutive positive integers is a multiple of 24.

Theorem. *The product of any four consecutive integers is a multiple of* 24.

Proof. Suppose that we multiply together four consecutive integers $n = abcd$. Since the integers follow the pattern even-odd-even-odd, we know that two of the four numbers must be even. Indeed, since every other even number is a multiple of 4, we know that one of those even numbers must be a multiple of 4. So n will have to be a multiple of 8. Similarly, at least one of the four numbers will be a multiple of 3, and so n also will be a multiple of 3. So n is a multiple of the least common multiple of 3 and 8, which is 24. ☐

Note that 24 is the best possible for this result, in light of the fact that $1 \cdot 2 \cdot 3 \cdot 4 = 24$. I should like to call attention, however, to something curious here. Namely, the number 24 happens to be the same as the factorial $4! = 24$. Does this suggest a more general result? Perhaps the product of any k consecutive integers is a multiple of $k!$? This is certainly true when the numbers start with 1, since the product would be $1 \cdot 2 \cdot 3 \cdots k$, which is exactly $k!$. Also, the result is true when $k = 1$, since every number is a multiple of $1! = 1$. Can we hope for the general result? Yes, indeed.

Theorem. *The product of any k consecutive integers is a multiple of $k!$.*

Proof. I claim that it suffices to consider only positive integers. The reason is that the product of consecutive negative integers will involve at most a difference of sign from the positive analogue, which won't matter for being a multiple of $k!$, and if both positive and negative integers appear amongst the consecutive integers, then 0 must also appear, and so the product will be 0, which is a multiple of $k!$. So we can restrict our attention to consecutive positive integers.

Suppose that we have a list of k consecutive positive integers

$$n+1 \quad n+2 \quad n+3 \quad \cdots \quad n+k.$$

The product is

$$(n + 1)(n + 2) \cdots (n + k).$$

We will prove by nested induction on n and k that this is a multiple of $k!$. (We discuss the method of induction further in chapter 4, but let me use it here for this extra result.) We have already observed that this is true in the trivial cases of $k = 1$ or when the sequence of numbers starts with 1. Let us assume that we have the sequence as shown, and that we have already established the result for all products of fewer than k consecutive integers and for products of k consecutive integers beginning earlier than $n + 1$. Using distributivity, we may expand the product on the last term, obtaining

$$(n + 1)(n + 2) \cdots (n + k - 1)(n + k) \quad =$$

$$n(n + 1)(n + 2) \cdots (n + k - 1) \quad + \quad k(n + 1)(n + 2) \cdots (n + k - 1).$$

The first summand of this result is the product of k consecutive numbers, starting one earlier than before, and so by our induction hypothesis, this is a multiple of $k!$. The second summand is k times the product of $k - 1$ consecutive integers, which will therefore be k times a multiple of $(k - 1)!$, and so the second summand also is a multiple of $k!$. So our product is the sum of two multiples of $k!$ and is therefore a multiple of $k!$ itself, as desired.   □

2.9 State and prove a theorem concerning positive integers k for which the product of any $k - 1$ consecutive positive integers is a multiple of k.

We can provide a complete solution—a necessary and sufficient condition for this property.

Theorem. *For any positive integer k, the product of any $k - 1$ consecutive positive integers is a multiple of k if and only if k is not prime and $k \neq 4$.*

Proof. Notice first that if k is prime, then $1 \cdot 2 \cdot 3 \cdots (k-1)$ is the product of $k-1$ consecutive positive integers that is not a multiple of k. It is not a multiple of k when k is prime, because none of the factors is a multiple of k, and so k cannot appear in the prime factorization of the product. Notice next that if $k = 4$, then $1 \cdot 2 \cdot 3 = 6$ is the product of three consecutive integers that is not a multiple of 4. Thus, we have established the forward implication by contraposition: if the product of any $k - 1$ consecutive integers is a multiple of k, then it must be that k is not prime and also $k \neq 4$.

For the converse implication, suppose that k is not prime and that $k \neq 4$. If $k = 1$, then the product of $k - 1$ many integers is the empty product, which is 1, and this is indeed a multiple of k in this case. Otherwise, since 2 and 3 are prime and $k \neq 4$, it follows that $4 < k$. Since k is not prime, we must have $k = rs$ for some positive integers $r, s < k$. Both r and s are at least 2, and since $4 < k$ it follows that one of them is at least 3. From this it follows that $r + s < rs = k$, because if $r < s$, then $r + s < 2s \leq rs = k$, and if $r = s$, then both are at least 3, and so $r + s = 2s < 3s \leq rs = k$. So the product of $k - 1$ many consecutive integers can be split into a product of r consecutive integers with the product of s consecutive integers, with whatever remains from the $k - 1$ integers. But by exercise 2.5, the product of the first r many of the integers is a multiple of r, and the product of the next s many is a multiple of s, and so the whole product will be a multiple of rs, which is k. So indeed, the product of any $k - 1$ consecutive integers will be a multiple of k, as desired.   □

3 Number Theory

Let us come to number theory, the queen of mathematics, a pure abstract distillation of mathematical thought. Here, we shall solve several problems concerning the prime numbers and features of the greatest common divisor of two numbers.

3.1 Prove that a positive integer is prime if and only if it has exactly two positive integer divisors.

Theorem. *A positive integer is prime if and only if it has exactly two positive integer divisors.*

Proof. This is an if-and-only-if statement, and we shall prove each direction separately. Assume that p is a positive integer.

($\rightarrow$) For the forward implication, suppose that p is prime. This means that p is an integer with $p > 1$ and the only positive divisors of p are 1 and itself. Since $p > 1$, these numbers are different, and so we see that p has exactly two positive integer divisors.

($\leftarrow$) Suppose now conversely that p is a positive integer with exactly two positive integer divisors. It follows that $p \neq 1$, since 1 has only one such divisor. So $p > 1$. And since both 1 and p are divisors of p, and these are different, they must be the only divisors of p. So p is prime. $\square$

3.2 Show that a positive integer $p > 1$ is prime if and only if, whenever p divides a product ab of integers, then either p divides a or p divides b.

Theorem. *A positive integer $p > 1$ is prime if and only if, whenever p divides a product ab of integers, then either p divides a or p divides b.*

Proof. This is an if-and-only-if statement, and so we prove each direction separately.

($\rightarrow$) This direction is precisely the content of lemma 13 in the main text.

($\leftarrow$) We prove the contrapositive. Assume that p is a positive integer, larger than 1, and is not prime. Thus, p has a positive integer divisor a other than 1 or p, which means $p = ab$ for some integers with $1 < a, b < p$. Thus, p divides ab, since it is equal to ab, but p does not divide a and does not divide b, since a larger positive integer cannot divide a smaller one. $\square$

3.3   Prove a stronger version of Bézout's lemma, namely, that for any two integers a and b, the smallest positive number d expressible as an integer linear combination of a and b is precisely the greatest common divisor of a and b.

Theorem.   *For any two integers a and b, the smallest positive number d expressible as an integer linear combination of a and b is precisely the greatest common divisor of a and b.*

Proof. We shall prove this theorem by generalizing the proof of the version of Bézout's lemma given in the main text. Consider any two integers a and b, and let d be the smallest positive integer that can be expressed as an integer linear combination of a and b, that is, as $d = ax + by$ for some integers x and y. Note that $d \leq a$, since $a = a \cdot 1 + b \cdot 0$, and that $d \leq b$, since $b = a \cdot 0 + b \cdot 1$. I claim first that d divides both a and b. To see this, apply the Euclidean division algorithm to express $a = kd + r$ for some integer k and remainder r, with $0 \leq r < d$. Putting these equations together, we observe that

$$r = a - kd = a - k(ax + by) = (1 - kx)a + (-ky)b.$$

We have therefore expressed r as an integer linear combination of a and b. Since $r < d$ and d was the smallest positive such combination, it follows that r must be 0. In other words, $a = kd$ is a multiple of d, as claimed. A similar argument shows that b also is a multiple of d, and so d is a common divisor of a and b. To see that d is the greatest common divisor, observe that any common divisor e of a and b will be a divisor of any integer linear combination $ax + by$ of a and b, since e divides each summand. So e must be a divisor of d, as d is such a linear combination. In particular, d must be at least as large as e, and so d is the greatest common divisor of a and b, as desired. $\square$

Let me draw out in detail what I find to be a surprising consequence of the argument we just made at the end of the proof.

Corollary.   *The greatest common divisor of two integers a and b is not merely the largest of the common divisors of a and b but is also a multiple of every common divisor of a and b.*

Proof. Let d be the greatest common divisor of integers a and b. The theorem shows that d can be represented as an integer linear combination $d = ax + by$ of the two numbers, and in fact it is the smallest positive integer linear combination of a and b. As we had observed at the end of the proof of the theorem, if e is a common divisor of a and b, then e divides $ax + by$, since it divides each summand, and so e divides d. In other words, d is a multiple of e.   □

Alternative second proof of the theorem. Let us now give a second proof of the theorem, proving it as a *consequence* of the version of Bézout's lemma given in the main text. According to that version of the result, if a and b are relatively prime, then we can express $1 = ax + by$ as an integer linear combination of a and b.

Consider now any two integers a and b, not necessarily relatively prime. Let us divide both a and b by their greatest common divisor d, whatever it is, obtaining the integer quotients $a' = a/d$ and $b' = b/d$. These must be relatively prime, for otherwise there would have been a greater common divisor. Therefore, by the relatively prime version of Bézout's lemma, it follows that $1 = a'x + b'y$ for some integers x and y. Multiplying through by d, we see that $d = a'dx + b'dy = ax + by$, thereby obtaining the greatest common divisor d as an integer linear combination of a and b. It must be the smallest positive such linear combination, since d divides any linear combination $ar + bs$, as it divides each term.   □

3.5 Show that if we count the number 1 as prime, then the uniqueness claim of the fundamental theorem of arithmetic would be false. (And this is one good reason not to count 1 as amongst the prime numbers.)

Theorem. *If the number 1 were included amongst the prime numbers, then no number would have a unique prime factorization. Indeed, every number would have infinitely many different prime factorizations.*

Proof. If 1 were counted as prime, then every number would have infinitely many different prime factorizations, since we could always add on more 1s as factors, as in

$$15 = 3 \times 5 = 3 \times 5 \times 1 = 3 \times 5 \times 1 \times 1 = 3 \times 5 \times 1 \times 1 \times 1.$$

The same idea works with any number n, since $n = n \times 1 = n \times 1 \times 1$, and so on.   □

This is one of the sound reasons not to count 1 as a prime number, for otherwise we would have to qualify the statement of the fundamental theorem of arithmetic, saying that every number had a unique factorization into *nonunitary* primes. Similar restrictions would be called for in many other contexts. It is simpler just to agree that 1 is not prime.

> **3.7** Mathematician Evelyn Lamb fondly notes of any large prime number presented to her that it is "one away from a multiple of 3!" And part of her point is that this is true whether you interpret the exclamation point as an exclamation or instead as a mathematical sign for the factorial. Prove that every prime larger than 3 is one away from a multiple of 3!.

Theorem. *Every prime number larger than 3 differs by one from a multiple of 6.*

Proof. Suppose that p is prime and larger than 3. Consider p modulo 6, which means we consider the remainder after dividing by 6. So $p = 6k + r$, where $0 \leq r < 6$. The remainder cannot be even, for then p would be even, which is impossible since p is prime and larger than 2. Similarly, the remainder cannot be 3, since then $p = 6k + 3$ would be a multiple of 3, which again is impossible since p is prime and larger than 3. So the remainder must be either 1 or 5, the only remaining remainders. But in this case, we either have $p = 6k + 1$, which differs by one from a multiple of 6, or we have $p = 6k + 5$, which differs by one from the next multiple of 6. So in any case, p is one away from a multiple of 6. □

Every multiple of 6 is also a multiple of 3, but the fun of Evelyn Lamb's way of saying it, of course, is that $3! = 6$.

> **3.9** Define that an integer is *even* if it is a multiple of 2; and otherwise it is *odd*. Show that every odd number has the form $2k + 1$ for some integer k. Conclude that any two consecutive integers consist of one even number and one odd number.

This definition of even and odd is different from the definition that we had used in chapter 1. The point here is that we can adopt this alternative definition, because the theorem we prove here shows in effect that the two definitions are equivalent.

Definition. An integer is *even* if it is a multiple of 2, and otherwise *odd*.

Theorem. *The even integers have the form 2k, for some integer k, and the odd integers have the form $2k + 1$.*

Proof. Consider an integer x and its remainder after division by 2. By the division algorithm, we find $x = 2k + r$ for some unique integer k and remainder integer r, with $0 \leq r < 2$. So r is either 0 or 1. If $r = 0$, then $x = 2k$, which is a multiple of 2 and hence even. If $r = 1$, then by the uniqueness of the result of the division algorithm, x is not even, and so it is odd, with the form $2k + 1$. □

Note the critical role played by the *uniqueness* of the remainder in the previous argument. When a number has the form $2k + 1$, then it is because of the uniqueness of the remainder that it cannot also be written in some other way as $2k'$, and this is why it is not even.

> **3.11** Criticize the following "proof." Claim. There are infinitely many primes. Proof. Consider any number n. Let $n!$ be the factorial of n, and consider the number $p = n! + 1$. So p is larger than n, and it has a remainder of 1 dividing by any number up to n. So p is prime, and we have therefore found a prime number above n. So there are infinitely many prime numbers. □

Let us criticize the proposed proof. It is not correct. Consider the claim that is being made.

"Claim." There are infinitely many primes.

This claim is true, and it can be proved by a correct argument, such as the argument given in the main text, an argument that in its essence is thousands of years old, going back at least to Euclid. But let us consider the argument offered here.

"Proof." Consider any number n.

The proof starts out basically fine. We want to prove there are infinitely many primes, and so we aim to construct arbitrarily large instances. By fixing the number n, we hope to construct a prime number larger than n. One could quibble at the use of *number* here, as opposed to *natural number* or *positive integer*, since there are many different number systems, such as complex numbers and ordinal numbers, but we are working here with the natural numbers.

Let $n!$ be the factorial of n, and consider the number $p = n! + 1$.

This part also seems fine, and it seems to be following the proof idea of Euclid's argument from the main text, which involved multiplying together all the primes on a given list, and then adding 1. Here, we are multiplying together all the numbers up to n, and then adding 1—a very similar strategy.

So p is larger than n, and it has a remainder of 1 dividing by any number up to n.

Yes, p is larger than n, because it is larger than $n!$, which is at least as large as n; and yes, it has a remainder of 1 when dividing by any (nontrivial) number up to n, since $n!$ is a multiple of all of those numbers, and the $+1$ forming p makes the remainder 1. One quibble here is that we should be talking about nontrivial divisors here, that is, larger than 1, since when dividing by 1 the remainder is 0.

So p is prime, and we have therefore found a prime number above n.

This is a critical error in the argument. All we've proved so far is that p does not have any nontrivial divisors up to n, but in order to be prime, we must know that it has no nontrivial divisors up to p, which could be considerably larger than n. In fact, we can make a counterexample to this step in the argument by considering the case $n = 4$, since in this case $p = 4! + 1 = 24 + 1 = 25$, which is not prime. Note that in this case, 25 does not have any nontrivial divisors up to 4, but it does have a nontrivial divisor above that, namely, 5.

So there are infinitely many prime numbers. □

If we had indeed managed to prove that p was prime, then this final step would be fine, since we would have found a prime number above any given number, and so in this case the primes would be unbounded in the natural numbers. So there would be infinitely many. But because of the earlier flaw in the argument, we are not entitled to this claim here on the basis of the argument provided.

> **3.13** Prove that every positive integer m can be factored as $m = rs^2$, where r is square-free.

Recall that a positive integer r is *square-free* when it is a not a multiple of any nontrivial square.

Theorem. *Every positive integer m can be factored as $m = rs^2$, where r is square-free.*

Proof. Consider any positive integer m, and let s be the largest integer for which s^2 divides m. We may write $m = rs^2$ for some integer r. If r were not square-free, then there would be a number $t > 1$ with t^2 dividing r. But in this case, we would have $r = kt^2$ for some k, and hence $m = rs^2 = kt^2s^2 = k(ts)^2$. So ts would be a number larger than s, whose square divides m. This contradicts the choice of s as the largest such number. So r must be square-free.   □

Alternative proof. Let m be any positive integer, and consider the prime factorization of m,

$$m = p_1^{k_1} p_2^{k_2} \cdots p_n^{k_n}.$$

Some of the exponents may be even, and some of them may be odd. Let r be the product of the primes that have an odd exponent in this prime factorization of m. Since r has no repeated prime factors, it follows that r is square-free. And once we have factored out one prime from each odd-exponent term, it follows that all the exponents in what is left will be even. Thus, that part can be written as s^2, where to form s we cut all the exponents in half. So $m = rs^2$, where r is square-free, as desired.   □

4 Mathematical Induction

Mathematical induction is a core principle of number theory, an engine of proof, a fundamental tool used to prove essentially all the foundational number-theoretic claims. The method of induction admits diverse equivalent formulations—common induction, strong induction, and the least-number principle, to name several—and different mathematical problems or situations often call for one or another of these forms over others. For this reason, I urge you to strive for a deep understanding of mathematical induction. Pick the right form of induction for your application.

4.1 Show by induction that $2^n < n!$ for all $n \geq 4$.

Theorem. $2^n < n!$ *for all integers* $n \geq 4$.

Proof. We shall prove the theorem by common induction. For the anchor case, consider $n = 4$, a case for which the claim is true, because $2^4 = 16$, which is less than $24 = 4!$. Now assume by induction that the claim is true at a number $n \geq 4$, so that $2^n < n!$, and consider $n + 1$. Let us multiply both sides of this inequality by 2, concluding $2^{n+1} < 2 \cdot n!$. Since $n + 1$ is larger than 2, we may continue $2 \cdot n! < (n + 1)n! = (n + 1)!$. Thus, $2^{n+1} < (n + 1)!$, and we have thereby established the claim for $n + 1$. So by induction, the claim is true for all integers $n \geq 4$. ☐

Note that the claim is not true when $n = 3$, since $2^3 = 8$ but $3! = 6$, and so $n \geq 4$ is the best possible hypothesis here.

4.3 Show by induction that $f_0 + \cdots + f_n = f_{n+2} - 1$ in the Fibonacci sequence.

We begin by setting up our theorem statement.

Theorem. *The Fibonacci sequence obeys the identity*

$$f_0 + \cdots + f_n = f_{n+2} - 1.$$

Proof. We shall prove the theorem by common induction on n. When $n = 0$, the claim is that $f_0 = f_2 - 1$, which is true because $f_0 = 0$ and $f_2 = 1$. Suppose now by induction that the identity holds for a natural number n, and consider $n + 1$. Our induction assumption is that $f_0 + \cdots + f_n = f_{n+2} - 1$ for this particular n. Let us add f_{n+1} to both sides of this equation, giving

$$f_0 + \cdots + f_n + f_{n+1} = f_{n+2} - 1 + f_{n+1} = (f_{n+1} + f_{n+2}) - 1.$$

Since $f_{n+1} + f_{n+2} = f_{n+3}$ according to the recursive Fibonacci rule, it follows that

$$f_0 + \cdots + f_{n+1} = f_{n+3} - 1,$$

which means that the identity holds for $n + 1$. So it holds by induction for all natural numbers n. $\qquad\qquad\square$

    Facts about the Fibonacci sequence can often be proved by induction, making use of the recursive definition of the sequence.

4.5 Consider an alternative Fibonacci sequence, starting with $0, 2$. Can you prove the analogue of theorem 25? Generalize the result as far as you can.

Theorem. *Consider the alternative Fibonacci sequence, proceeding*

$$0 \quad 2 \quad 2 \quad 4 \quad 6 \quad 10 \quad 16 \quad \cdots$$

defined by the recursion $x_0 = 0$, $x_1 = 2$, and $x_{n+2} = x_n + x_{n+1}$. In this case,

$$x_0^2 + x_1^2 + \cdots + x_n^2 = x_n x_{n+1}$$

for every natural number n.

Proof. We prove the theorem by common induction. The theorem holds in the case $n = 0$, because $x_0^2 = 0$, which is equal to $x_0 x_1 = 0 \cdot 2$. Suppose by induction that the identity holds for a particular value of n, so that

$$x_0^2 + x_1^2 + \cdots + x_n^2 = x_n x_{n+1}.$$

We consider $n + 1$. By adding x_{n+1}^2 to both sides, we may deduce

$$x_0^2 + x_1^2 + \cdots + x_n^2 + x_{n+1}^2 = x_n x_{n+1} + x_{n+1}^2.$$

The right-hand side of this is can be factored as $(x_n + x_{n+1})x_{n+1}$, which is equal to $x_{n+2}x_{n+1}$. Therefore, we conclude that

$$x_0^2 + x_1^2 + \cdots + x_{n+1}^2 = x_{n+1}x_{n+2},$$

and so the claim holds for $n + 1$. Therefore, by induction, the claim holds for every natural number n. □

An alternative argument, for that particular sequence, could observe that the sequence of values are exactly double the usual Fibonacci sequence $x_n = 2f_n$. And since doubling every term will end up multiplying both sides of the identity by 4, we will still have equality.

But the inductive argument did not actually use much about the particular values of 0 and 2, and we can generalize the result as follows.

Theorem. *Suppose that x_0, x_1, x_2, ... is a sequence obeying the Fibonacci recursion $x_n + x_{n+1} = x_{n+2}$, but possibly with alternative starting values x_0 and x_1. If either $x_0 = 0$ or $x_0 = x_1$, then*

$$x_0^2 + x_1^2 + \cdots + x_n^2 = x_n x_{n+1}$$

for every natural number n.

Proof. Suppose that our sequence x_0, x_1, x_2, ... obeys the hypotheses. The extra assumption on x_0 and x_1 ensures exactly that $x_0^2 = x_0 x_1$, which was exactly what was needed in the induction proof above. And the induction step part of that proof used only the recursion rule $x_n + x_{n+1} = x_{n+2}$, which still holds in this case. So by induction, we may conclude that

$$x_0^2 + x_1^2 + \cdots + x_n^2 = x_n x_{n+1}$$

for every natural number n. □

4.7 Show by induction that a finite set with n elements has exactly 2^n many subsets.

Theorem. *A finite set with n elements has exactly 2^n many subsets.*

Proof. We prove the theorem by common induction. The claim is certainly true when $n = 0$, for the empty set ∅ has exactly one subset, namely, ∅ itself, and $2^0 = 1$. Suppose that the claim is true for some particular value of n, and consider $n + 1$. Suppose that A is a set with $n + 1$ elements. Let us pick a particular element $a \in A$. We shall count the subsets of A in two groups: those that do not contain the point a, and those that do. The subsets of A that do not contain a are exactly the subsets of $A \setminus \{a\}$, which is a set of size n. By the induction hypothesis, there are exactly 2^n many such subsets. And to each of these subsets, we can add the point a, thereby forming exactly the subsets of A that do contain the point *a.* So we have exactly 2^n many subsets of A that omit a and exactly 2^n many subsets that contain a, yielding exactly $2^n + 2^n = 2^{n+1}$ many subsets of A in all. This verifies the claim for $n + 1$. So by induction, every finite set of size n has exactly 2^n many subsets. □

4.9 Prove that $\frac{1}{1\cdot2} + \frac{1}{2\cdot3} + \frac{1}{3\cdot4} + \cdots + \frac{1}{n(n+1)} = \frac{n}{n+1}$.

Theorem. *For every positive integer n,*

$$\frac{1}{1\cdot2} + \frac{1}{2\cdot3} + \frac{1}{3\cdot4} + \cdots + \frac{1}{n(n+1)} = \frac{n}{n+1}.$$

Proof. We prove the identity by common induction. The claim is true when $n = 1$, because $1/(1 \cdot 2) = 1/2$. Suppose that the identity holds for a particular value of n, and consider $n + 1$. Let us add the next term $1/(n + 1)(n + 2)$ to both sides, obtaining

$$\frac{1}{1\cdot2} + \frac{1}{2\cdot3} + \frac{1}{3\cdot4} + \cdots + \frac{1}{n(n+1)} + \frac{1}{(n+1)(n+2)} = \frac{n}{n+1} + \frac{1}{(n+1)(n+2)}.$$

The right-hand side can be simplified by finding a common denominator as follows:

$$\frac{n}{n+1} + \frac{1}{(n+1)(n+2)} = \frac{n(n+2)+1}{(n+1)(n+2)} = \frac{n^2+2n+1}{(n+1)(n+2)} = \frac{(n+1)^2}{(n+1)(n+2)} = \frac{n+1}{n+2}.$$

This establishes the identity for $n + 1$. And so by induction, the claim holds for all positive integers n. □

4.11 Show that every natural number has a unique base 3 representation.

Theorem. *Every positive integer has a unique base 3 representation, that is, as a sum of the form*

$$d_0 + d_1 3 + d_2 3^2 + \cdots + d_k 3^k,$$

where each base 3 digit d_i is either 0, 1, or 2 and the final term is nonzero $d_k \neq 0$.

Proof. The theorem is making two claims, an existence claim and a uniqueness claim. Let us prove them both simultaneously by strong induction. Assume that the claim is true for all numbers smaller than n. Choose some largest 3^k fitting inside n, so that $3^k \leq n$. If $3^k = n$, then $n = 1 \cdot 3^k$ provides a base 3 representation. So assume that $3^k < n$. The number $n - 3^k$, therefore, is a positive integer smaller than n, and so by the induction assumption, it has a unique base 3 representation. Note that we cannot have $2 \cdot 3^k$ in the representation of $n - 3^k$, for then we would have $3 \cdot 3^k \le n$ and hence $3^{k+1} \le n$, contrary to the choice of k. So the 3^k term in $n - 3^k$ is either $0 \cdot 3^k$ or $1 \cdot 3^k$. We may therefore simply add 3^k, making this term either $1 \cdot 3^k$ or $2 \cdot 3^k$, and thereby produce a base 3 representation of n, as desired. This proves existence. For uniqueness, let us argue in analogy with the argument of the main text for binary sequences, namely, that 3^k must appear in any base 3 representation of n. For this, we use the following lemma, asserting that $2 + 2 \cdot 3 + 2 \cdot 3^2 + \cdots + 2 \cdot 3^{k-1} = 3^k - 1$, and so even if one used all the smaller terms, it wouldn't be enough. Thus, the uniqueness of the representation of n follows from the uniqueness of the base 3 representation of $n - 3^k$. $\square$

Lemma. *For any natural number k,*

$$2 + 2 \cdot 3 + 2 \cdot 3^2 + \cdots + 2 \cdot 3^k = 3^{k+1} - 1.$$

Proof. We prove the identity by common induction. This is true for $k = 0$, since $2 = 3^1 - 1$. Assume by induction that the identity is true for a particular number k, and consider $k + 1$. By adding $2 \cdot 3^{k+1}$ to both sides of the equation, we see that

$$2 + 2 \cdot 3 + 2 \cdot 3^2 + \cdots + 2 \cdot 3^k + 2 \cdot 3^{k+1} = 3^{k+1} - 1 + 2 \cdot 3^{k+1}.$$

The right-hand side simplifies to

$$3^{k+1} - 1 + 2 \cdot 3^{k+1} = 3 \cdot 3^{k+1} - 1 = 3^{k+2} - 1,$$

which establishes the identity for $k + 1$. So by induction, the claim holds for all natural numbers. $\square$

4.12 Prove that if you divide the plane into regions using finitely many straight lines, then the regions can be colored with two colors in such a way that adjacent regions have opposite colors.

Theorem. *If we divide the plane into regions using finitely many straight lines, then the regions can be colored with two colors in such a way that adjacent regions have opposite colors.*

Proof. We prove the theorem by induction on the number of lines. We might begin with the case of zero lines, in which case we may color the entire plane with just one color. Or, if you like to interpret the problem as having at least one line, then we may anchor the induction with one line, which divides the plane into two half-plane regions, and we may color them with different colors.

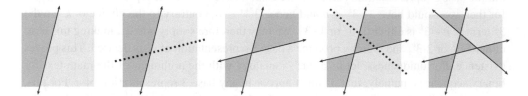

Suppose that we have placed *n* lines in the plane and colored the regions with two colors in such a way that fulfills the requirements. Place a new line ℓ in the plane. The line ℓ divides the plane into two half planes. Let us simply swap the colors one for the other, but only on one side of ℓ, while keeping the original colors on the other side of ℓ. In this way, the regions immediately cut by ℓ will exhibit the adjacent-regions-opposite-color feature, and this will also remain true for any other line, since the feature held originally, and it is preserved when you swap colors. Note that it is no problem if the new line should happen to pass through an intersection of previously occurring lines.   □

An alternative proof proceeds like this: Put all the lines in the plane from the start, and then color each region with one color or the other, depending on whether one must respectively cross an even or odd number of lines altogether to get to that region from some fixed point not on any of the lines. This parity count is well defined, independently of the particular path of getting to that region, provided that the lines are always crossed transversely from one side to the other and that there are only finitely many such crossings. And adjacent regions will differ by exactly one in the parity count, and hence they will get opposite colors.

4.13 True or false: every flying polka-dotted elephant in this room is smoking a cigar.

The statement "every flying polka-dotted elephant in this room is smoking a cigar" is true, considering the room I currently occupy. Indeed, for my room it is vacuously true, because there are no elephants in my room at all, and certainly no polka-dotted elephants, or any flying polka-dotted elephants. And so it is true that every one of them—which is none—is smoking a cigar, because it is not possible to exhibit a counterexample to this, which would be a particular such elephant that is *not* smoking a cigar. Since one cannot exhibit an elephant in my room at all, one certainly cannot exhibit a flying polka-dotted elephant that is not smoking a cigar. And since there are no counterexamples to the statement, it is true.

Notice that vacuous truth applies only to universal as opposed to existential statements; the statement was about *every* such-and-such kind of elephant; since there were no such elephants, anything that we might have said about them would be true, vacuously true. Meanwhile, the related existential statement that "there is a flying polka-dotted elephant in this room that is smoking a cigar" is not true.

I would often tell my seven-year-old-daughter that she was my absolutely favorite seven-year-old daughter, who went to such-and-such a school and who lived at such-and-such an address. Knowing about vacuous truth, she would reply, "But Papa, aren't I also your least favorite such person?" And I had to admit to her that, yes indeed, that also was true, but I would never say it that way myself.

4.15 Criticize the following "proof." Claim. $\sum_{i=1}^{\infty} \frac{1}{i} < \infty$. Proof. Let $S(n) = \sum_{i=1}^{n} \frac{1}{i}$, and consider the statement asserting $S(n) < \infty$. This statement is true for $n = 1$, since $\frac{1}{1} = 1 < \infty$. And if it is true for n, then it is true for $n + 1$, since the next sum $S(n+1) = \sum_{i=1}^{n+1} \frac{1}{i}$ is equal to the previous sum $\sum_{i=1}^{n} \frac{1}{i}$ plus the next term $\frac{1}{n+1}$, or in other words, $S(n+1) = S(n) + \frac{1}{n+1}$, and this is the sum of two finite numbers. So $S(1)$ is true, and $S(n)$ implies $S(n+1)$. So we have therefore proved $\sum_{i=1}^{\infty} \frac{1}{i} < \infty$ by induction. $\square$

Let us consider the argument.

> **"Claim."** $\sum_{i=1}^{\infty} \frac{1}{i} < \infty$.

The claim is not true, since this is the famous harmonic series, which is well known to diverge despite passing the test for divergence (see the subsequent theorem). But we may continue to analyze this particular argument to see what problems may arise.

> **"Proof."** *Let $S(n) = \sum_{i=1}^{n} \frac{1}{i}$, and consider the statement asserting $S(n) < \infty$.*

We seem to be setting up for a proof by induction. The statement that $S(n) < \infty$, however, is clearly true for every n, since $\sum_{i=1}^{n} \frac{1}{i}$ is a finite sum of real numbers, and so each $S(n)$ is obviously finite. So there is something a bit strange about this argument already.

> *This statement is true for $n = 1$, since $\frac{1}{1} = 1 < \infty$. And if it is true for n, then it is true for $n + 1$, since the next sum $S(n+1) = \sum_{i=1}^{n+1} \frac{1}{i}$ is equal to the previous sum $\sum_{i=1}^{n} \frac{1}{i}$ plus the next term $\frac{1}{n+1}$, or in other words, $S(n+1) = S(n) + \frac{1}{n+1}$, and this is the sum of two finite numbers.*

This is a perfectly sound proof by induction that $S(n) < \infty$ for every natural number n.

> *So $S(1)$ is true and $S(n)$ implies $S(n+1)$.*

This statement does not make sense, since $S(1)$ is a number rather than a statement. What had been proved is the statement "$S(n) < \infty$," and this latter statement is what is true at $n = 1$ and whose truth at n implies its truth at $n + 1$.

So we have therefore proved $\sum_{i=1}^{\infty} \frac{1}{i} < \infty$ by induction. □

This final conclusion is just not right. At no point in the induction proof did we consider the *infinite* sum. Every step was concerned merely with having one more term in a strictly finite sum. The conclusion of the inductive argument is that $\sum_{i=1}^{n} \frac{1}{i}$ is finite for every natural number *n*. But this is not at all the same as saying $\sum_{i=1}^{\infty} \frac{1}{i}$ is finite. For example, every finite sum $1 + 1 + 1 + \cdots + 1$ is finite, but this doesn't mean an infinite sum of 1s is finite.

Theorem. *The harmonic series $\sum_{i=1}^{\infty} \frac{1}{i}$ diverges to infinity.*

Proof. Consider the finite partial sum $S(n) = \sum_{i=1}^{n} \frac{1}{i}$. Let's double the number of terms by moving to $S(2n) = \sum_{i=1}^{2n} \frac{1}{i}$. The difference between $S(n)$ and $S(2n)$ consists precisely of the extra terms

$$\frac{1}{n+1} + \frac{1}{n+2} + \cdots + \frac{1}{2n}.$$

The terms are getting smaller—the smallest is the last one—and we have *n* extra terms here in all. So the sum of these extra terms is at least $n \cdot \frac{1}{2n}$, which is $1/2$. Therefore, by moving from $S(n)$ to $S(2n)$ we have added at least $1/2$ to the total sum. So we can add as much as we like to the total sum, exceeding any given finite quantity, simply by adding it $1/2$ at a time, by often-enough doubling the number of terms in a given finite partial sum. Therefore, $\sum_{i=1}^{\infty} \frac{1}{i}$ is infinite, and so the series diverges to infinity.   □

> **4.17** Prove that the least-number principle, the common induction principle, and the strong induction principle are all equivalent. Assuming any one of them, you can prove the others.

Theorem. *The following principles are equivalent.*

1. *The least-number principle: Every nonempty set of natural numbers has a least element.*

2. *The common induction principle: If A is a set of natural numbers with $0 \in A$ and for which $n \in A$ implies $n + 1 \in A$, for every n, then A contains every natural number.*

3. *The strong induction principle: If A is a set of natural numbers and $n \in A$ whenever every $k < n$ is in A, then every natural number is in A.*

But what is going on here? Any two *true* statements are logically equivalent, and so if we have assumed that all the various forms of induction are true, then they are trivially equivalent to one another. Is the theorem trivial? No, that is not the sense in which in the theorem should be taken. Rather, what we are claiming is that the three induction principles are equivalent, *without assuming any of them are true*. That is, we shall prove over a weak theory, without induction, that the various forms of induction are equivalent. Of course, when doing number theory and undertaking the mathematical development of number theory, we shall assume the induction principle as an axiom, and the point of the theorem above is that in that undertaking it doesn't matter which particular form of induction we take as an axiom, precisely because they are all equivalent. Any number system satisfying the least-number principle will also satisfy the common induction principle and the strong induction principle and vice versa in the converse implications.

Proof. (1 → 2) Assume the least-number principle, and suppose that A is a set of numbers satisfying the hypotheses of the common induction principle. If there is some natural number that is not in A, then by the least-number principle, there must be a smallest number m that is not in A. This number cannot be 0, since we are given that $0 \in A$. So it must have the form $m = n + 1$. Since m was the smallest number not in A, it follows that $n \in A$. But from this, by our induction assumption about A, it follows that $n + 1 \in A$ after all, a contradiction. So every number must be in A, and we have verified the common induction principle.

(2 → 3) Assume the common induction principle, and suppose that A is a set of numbers satisfying the strong induction property, namely, that $n \in A$ whenever every number smaller than n is in A. Let us prove by common induction that, for every number n, every number below n is in A. This is vacuously true for $n = 0$, since there are no natural numbers below 0. Suppose it is true for n. By the strong induction assumption we made about A, this implies $n \in A$, and therefore every number below $n + 1$ is in A. So by common induction, we've proved that for every number n, every number below n is in A. This implies every number is in A, since n is below $n + 1$.

(3 → 1) Assume the strong induction principle, and suppose that A is set of natural numbers without a least element. Let $B = \mathbb{N} \setminus A$ be the complement of A. Note that for any number n, if all smaller numbers than n are in B, then $n \in B$, for otherwise n would be a smallest element of A. So by the strong induction principle, it follows that every natural number is in B. In this case, A is empty. So we've proved that every nonempty set of natural numbers must have a least element. □

5 Discrete Mathematics

In this chapter, we explore several fun exercises in discrete mathematics concerning finite combinatorics, various game strategies, and tiling problems.

> **5.1** Suppose that a finite group of people has some pattern of pointing at each other, with each person pointing at some or all or none of the others or themselves. Prove that if there is a person who is more often pointed at than pointing, then there is another person who is less often pointed at than pointing.

Theorem. *In any finite group of people with a pattern of pointing at each other, with each person pointing at some or all or none of the others or themselves, if there is a person who is more often pointed at than pointing, then there is another person who is less often pointed at than pointing.*

Proof. Each instance of pointing is also an instance of being pointed at. Because of this, the sum of all the pointing scores for every individual is the same as the sum of the pointed-at scores over all individuals, for in both cases this is the total number of instances of pointing. If some individual is more often pointed at than pointing, then this individual deficit would cause an overall deficiency in pointing, unless it is made up by some other individuals. So there must be some other individuals who are less often pointed at than pointing.

More specifically, if person i has p_i instances of pointing and a_i instances of being pointed at, then the main observation is that, because every instance of pointing contributes once on each side of the finger, the two sums are equal:

$$p_1 + p_2 + \cdots + p_n = a_1 + a_2 + \cdots + a_n.$$

If there is some i with $p_i < a_i$, therefore, then there must be some j with $p_j > a_j$, for otherwise the left-hand sum would be too small to make the sums equal. □

> **5.3** Show that if there are infinitely many people, then it could be possible for every
> person to be more pointed at than pointing. Indeed, can you arrange infinitely
> many people, such that each person points at only one person but is pointed at by
> infinitely many people? How does this situation interact with the money-making
> third proof of theorem 34?

Let us imagine that the people are assigned distinct natural numbers,

$$0 \quad 1 \quad 2 \quad 3 \quad 4 \quad \cdots,$$

and so we might think of pointing relations on the set of all natural numbers.

Theorem. *There is a pointing relation on the natural numbers such that every number points at only one other number but is pointed at by infinitely many numbers.*

Proof #1. We begin by directing all the even numbers to point at 0. Next, amongst the numbers remaining, which are the odd numbers, we direct every other one of them to point at 1. And amongst the numbers remaining after this, we direct that every other one should point at 2, and so on. In this way, it is clear that every number points at just one number, but meanwhile every number will eventually have its turn for adulation, becoming at that time pointed at by infinitely many numbers.   □

Proof #2. Let us provide a more explicit pointing relation. Every positive integer has the form $n = 2^r(2k+1)$, since we may factor the largest power of 2 out of the prime factorization of n, and what remains is odd. This form of representing n is unique, by the uniqueness of the prime factorization.

Let us direct number n to point at the exponent number r. (And let us direct 0 to point at itself.) Thus, each number points at exactly one number, but meanwhile, every number r is pointed at by all the numbers of the form $2^r(2k+1)$, of which there are infinitely many.   □

How does this theorem relate to the money-making proof of theorem 34? Well, if we had infinitely many people, each with exactly one dollar, and we directed them to pay their dollar according to the pointing scheme of the theorem, then after the payments, every person would have infinitely many dollars! This may seem paradoxical at first, since each person seems to have made a lot of money, but where did the extra cash come from? We just moved the same money around within the group. The paradox may be dispelled when one realizes that the total number of dollars held by the group has not changed—it is still the same countable infinity. Thus, although from each individual person's perspective it may seem as though the group has made money, in fact the number of dollar bills has remained unchanged.

5.5 Generalize the chocolate bar theorem (theorem 36) to nonrectangular chocolate bars.

Theorem. *Suppose a chocolate treat consists of an assemblage of small squares in a grid pattern, not necessarily forming a rectangle, but forming a single connected shape. If one systematically breaks the chocolate treat along line segments formed across the treat by the grid lines, using the account in the main text of what it means to break a piece of chocolate, then it will always take the same number of steps to break the treat into individual squares regardless of the breaking protocol that is followed.*

Proof. The main text explains that to break a piece of chocolate means to take a single connected piece of chocolate and separate it into two nonempty pieces by cutting along one of the grid lines. With this understanding of what it means to break the chocolate, every break will cause one "piece" of chocolate to become two. And so the number of steps that it will take to separate all the squares is exactly one less than the total number of squares, since this is the number of pieces that one will have at the end. □

Note that one must pay careful attention to the definition of what it means to "break" a piece of chocolate, in light of chocolate figures shaped like this:

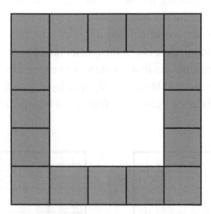

According to the explanation, it is allowed as an initial move to break along a long vertical line that passes through the whole chocolate figure, across the top part, through the gap in the center and also across the bottom part, since this would take the one chocolate ring piece and make two half-ring pieces. But after having done that, it would not be allowed to break along a similar horizontal line passing through two already-separated pieces, since that would make two pieces into four, whereas the official description said that a break should always increase the total number of pieces by exactly one.

5.7 Explain how the proof of theorem 37 provides a construction method for producing the desired tiling. (Namely, once a given square is omitted, then perform the division into quadrants, place the one tile, and iterate with the smaller squares.) What tiling do you get this way for the 16×16 grid shown in the illustration?

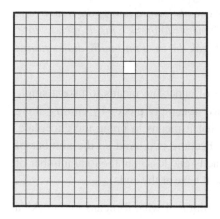

The proof of theorem 37 provides a general construction method for tiling a square board of size $2^n \times 2^n$, with one unit square omitted, using L-shaped tiles. Specifically, the inductive argument of the proof instructs us to divide the board into four smaller boards and place the first tile near the center, so that it covers one square from each of the four smaller boards other than the one containing the omitted square. This reduces to the tiling problem to four smaller instances, which now in effect each have one square omitted. The point is that we may simply carry on by dividing each of those boards into four and placing the next four tiles near the centers of them, so as to cover one square each, except for the one already having a square covered. The result after n iterations will be a tiling of the original board.

Let us illustrate the method in the case of the particular 16×16 grid used in the main text, shown here. We begin by placing the first tile near the center, arranged so that it covers one square on each of the other three subboards. We now have four 8×8 boards, each missing one square. So we may subdivide again, into four 4×4 boards, four times, and place a tile for each of them.

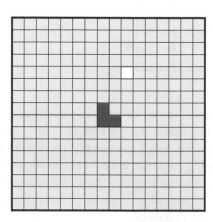
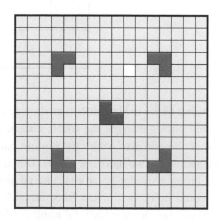

After dividing again into 2×2 boards, we achieve the full tiling.

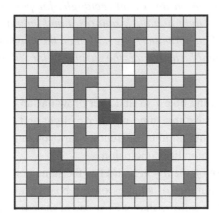 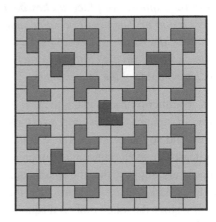

In this way, we systematically reduce to smaller instances, placing 4^k tiles at stage k, until we have reduced to 2×2 boards with one square omitted, which are exactly filled by a tile. The inductive proof provides a construction procedure for producing the tiling.

5.9   Prove or refute the following statement: If you place a chessboard pattern on an $n \times m$ rectangular grid, there will be an equal number of dark and light squares. Generalize your answer as much as you can.

We are considering the statement asserting that an $n \times m$ chessboard pattern always has equal number of light and dark squares. The statement is not true, and there are many counterexamples. For example, the 3×5 board pictured here has eight light squares and only seven dark squares. Similar examples can be made with the 3×3 board, the 5×7 board, and many others. What does it take to be a counterexample?

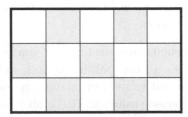

The smallest counterexample, of course, is a 1×1 board consisting of a single square:

Let us try to find necessary and sufficient conditions for equality.

Theorem. *Suppose we place a chessboard pattern on an n × m rectangle, for positive integers n and m. There will be equal numbers of light and dark squares if and only if at least one of the numbers is even. Furthermore, if both numbers are odd, then the numbers of light and dark squares differ by exactly one, and furthermore, the surplus color will be the common color of the four corners.*

Proof. If either n or m is even, then let us orient the board so that every row has an even number of squares. In this case, every row will have the same number of light and dark squares, and so the board overall will have the same number of light as dark squares. So the at-least-one-even criterion is sufficient for the equality conclusion.

Conversely, suppose that the chessboard pattern on the $n \times m$ board has the same number of light as dark squares. In particular, the total number of squares on the board must be even. But since the total number of squares is nm, it follows that at least one of the numbers n or m must be even. So the at-least-one-even criterion is also necessary for the equality conclusion.

Let us now consider the case that both numbers are odd. Let us imagine that we might cut off the last column of the board. What remains would have an even number of columns, and so by what we've proved above, that part of the board would have an equal number of light and dark squares. And on the cutoff column itself, we alternate light, dark, and so on, but because it is odd, we'd begin and end with the same color, and so have exactly one extra square of that color. Those beginning and ending squares of that column would be corner squares of the original board, and because the number on each row is odd, these will also be the common color of the other two corners.   □

5.11 Give an alternative proof of theorem 41 by induction.

Theorem. *Every positive integer n can be expressed as a sum of one or more positive integers in precisely 2^{n-1} many ways.*

Proof. Let us use the same notion of representing-as-a-sum as in the main text, where we may write n as a sum $k_1 + k_2 + \cdots + k_r$ of positive integers, without parentheses; the order of the summands matters here, and we allow trivial sums, consisting of just one summand. Let us prove the theorem by common induction. The number 1 has only one representation, as the number 1 itself, and so since $2^0 = 1$ the theorem is true for $n = 1$. Suppose that the theorem is true for a particular number n, and consider $n + 1$. Representations of $n + 1$ can arise by modifying representations of n.

For example, given any representation of n as

$$n = k_1 + \cdots + k_r,$$

we may form representations of $n + 1$ in two ways, either by adding a new term 1 on the end, as in

$$n + 1 = k_1 + k_2 + \cdots + k_r + 1,$$

or by adding 1 to the last term, forming

$$n + 1 = k_1 + k_2 + \cdots + \bar{k}_r,$$

where $\bar{k}_r = k_r + 1$, but all the other terms are kept the same. Every representation of $n + 1$ arises in one of these two ways from a representation of n, since if the representation of $n + 1$ ends with $\cdots + 1$, then it has the first form, but if it ends with a summand at least 2, then it has the second form. So these two forms are both exhaustive and have no common instances. The number of representations of $n + 1$, therefore, is exactly double the number for n, which by our induction assumption is exactly 2^{n-1}. So we have $2 \cdot 2^{n-1} = 2^n = 2^{(n+1)-1}$ many ways to represent $n + 1$, thereby showing that the statement is true for $n + 1$. Thus, by induction, the statement is true for every natural number. $\square$

5.13 Suppose one plays a version of the Escape! game on a finite $n \times n$ board, modifying the rules so that stones that would have been placed outside this region are simply not placed. For which values of n can you vacate the yellow corner? For which values of n can you vacate the entire board?

Theorem. *For any natural number n, consider the finite version of Escape!, played on a finite n × n board, with the rule that stones that would have been placed outside this region simply fall off the board, removed from play. In this version of the game, the player can play so as to vacate the shaded corner region. Indeed, every sufficiently long sequence of legal moves must inevitably vacate the entire board completely.*

Proof. The claim is that every sequence of legal moves, of sufficient length, will completely vacate the board. Notice first that if there are stones on the board, then there is always some legal move, since the upper-rightmost stone—meaning the top stone on the rightmost occupied column—can always be moved. In order to prove the theorem, we shall prove that there is no infinite sequence of legal moves. Thus, the board must become empty at some point.

Let us prove this claim by induction on the board size n. The reader may observe that the claim is true on the 2×2 board, the smallest board containing the original three stones, because (as the reader can verify) the 2×2 board must become empty after exactly nine moves. Assume the claim is true for the $n \times n$ board, and consider now the $(n + 1) \times (n + 1)$ board. Notice that making a move on an edge stone, meaning a stone on the upper edge or the right edge, will never cause a stone to be returned to the lower $n \times n$ part of the board. And furthermore, one can make only finitely many such edge moves consecutively, since each such move causes the stones to move closer to the upper right corner, without introducing any new edge stones. So if there is an infinite play, it will involve infinitely many moves on the lower $n \times n$ part of the board. But by the induction hypothesis, one can make only finitely many such moves on that part, after which it becomes completely vacated. So on the $(n + 1) \times (n + 1)$ board, eventually the only stones available will be edge stones, and then after finitely many additional moves, these will all also be vacated. So any sufficiently long play will vacate the board.   □

Alternative proof. Let us give an alternative proof. Consider the labeling of the board squares with the reciprocal powers of 2, as in the proof of theorem 39. In that proof, we had observed that the initial three stones had a total value of 1 and that moves in the original Escape! game preserve this value. In the modified game this remains true, provided that stones are not removed from play. But when a stone on the edge is moved, meaning the upper and right edges, then one or both of the new stones may fall off the board and be removed from play; in this case, the total value will go down. By how much? On an $n \times n$ board, every square has value at least $1/2^{2n-1}$, and so each time a stone is lost off the board, it will have value at least $1/2^{2n}$. Therefore, this can happen at most 2^{2n} many times, finitely many.

But I claim that any sufficiently long play must inevitably lead to a move on the upper or right edge, causing a stone to be removed from play. To prove this, let us consider the total integer *coordinate value* of an arrangement of stones, which is the sum total of all the x- and y-coordinates of the occupied squares, using $(0, 0)$ for the lower left, and then $(1, 0)$ and $(0, 1)$, and so on. Notice that every legal move, except the move of removing the upper-right corner stone, causes the total coordinate value strictly to increase. Because the total coordinate value of an arrangement of stones is bounded by the sum of the coordinate values of the entire board, the coordinate value cannot go up infinitely often. Therefore, in any sufficiently long play, there must be another instance where the upper-right corner stone is removed, since this is the only way for the total coordinate value to go down. But we have already argued that the upper-right corner stone can only be removed finitely many times. Therefore, every sequence of play of legal moves must be finite.   □

5.15 (Challenge) Show that the number of steps to vacate the $n \times n$ board does not depend on the particular sequence of moves that are made. The game always ends with an empty board in exactly the same number of steps regardless of how the plays are made.

Theorem. *In the $n \times n$ version of the Escape! game, every play leading to a vacated board takes exactly the same number of steps.*

Proof. Consider the $n \times n$ game. We've proved already that every sufficiently long play will vacate the board. I claim that every cell on the board will be activated or moved upon exactly the same number of times in any such vacating play, regardless of the order of play. Let us prove this by induction with respect to the relation of below-or-to-the-left. Clearly, the origin square at the lower left can be moved upon only once, since once it is played upon, it will never get another stone on it to be played again. Consider now another square on the board, and assume by induction that the squares below or to the left, if any, are played upon the same number of times in any vacating play, regardless of the order of play. Each such time, a new stone appears in our given square, and in order to vacate the board, it will have to be played upon. So the number of times we must play on our given square is simply the sum of the numbers from the square below and the square to the left, and the order of play does not matter. □

One can use the idea of the proof to calculate exactly how many moves it will take to vacate the $n \times n$ board. Specifically, let $m_{i,j}$ be the number of times that square (i, j) will be acted upon in a vacating play. The previous proof observes that $m_{0,0} = 1$, and one can similarly observe that $m_{1,0} = m_{0,1} = 2$, since those squares will each have to be moved upon to clear the starting configuration stone, and also to clear the stone they will inherit from the original corner stone. The inductive argument shows that $m_{0,j+1} = 2$ along the bottom edge and $m_{i+1,0} = 2$ along the left edge. And in general, the numbers will obey the

2	10	30	70	140
2	8	20	40	70
2	6	12	20	30
2	4	6	8	10
1	2	2	2	2

recursion $m_{i+1,j+1} = m_{i,j+1} + m_{i+1,j}$, meaning that the cell value is the sum from below and to the left. The total numbers can therefore be computed recursively.

By adding up all these numbers, we see that on the 5 × 5 board, a play is vacating if and only if it has exactly 501 moves. One can see the array of numbers as arising from the sum of three slightly displaced copies of Pascal's triangle, a separate copy starting at each of the three starting stones. I believe that this idea could eventually lead one to a closed-form expression for the total number of moves.

6 Proofs without Words

In the genre of proof known as *proof without words*, a mathematician attempts to convey the essence of a mathematical argument by means of a visual figure alone, without using any words. These proofs are often extremely clever, and it is amazing how fully a mathematical figure can convey a mathematical argument. Nevertheless, my view is that nearly every proof without words is improved by a few well-chosen words, and there is no special value or virtue in giving a proof truly without any words. Furthermore, too often in my experience, proofs offered as a proof without words are instead merely poorly explained proofs. So let us abandon any dogmatic approach to proofs without words and view the genre instead as a celebration of the power of a mathematical figure to convey mathematical ideas. Use insightful diagrams and figures in your proofs, yes, but kindly also use words to convey your meaning.

6.1   Give an alternative proof by induction that $1 + 2 + \cdots + n = \binom{n+1}{2}$.

The main text provided a proof without words; here, we give a proof without any figure.

Theorem. *The sum of the first n positive integers obeys the identity*

$$1 + 2 + \cdots + n = \binom{n + 1}{2}.$$

Proof. We prove the identity by common induction. The statement is true when $n = 1$, since $\binom{2}{2} = 1$, as there is only one way to choose 2 out of 2 items. Suppose that the identity is true for a number n, and consider $n + 1$. By adding the next term, we see that

$$1 + 2 + \cdots + n + (n + 1) = \binom{n + 1}{2} + (n + 1).$$

The right-hand side is equal to $\binom{n+2}{2}$, because to choose 2 elements from a set of size $n + 2$ means either to choose them from the first $n + 1$ elements, and there are $\binom{n+1}{2}$ ways to do that, or to use the very last point and pair it with one of the earlier $n + 1$ points, and there are $n + 1$ ways to do that. So we've established that

$$1 + 2 + \cdots + n + (n + 1) = \binom{n + 2}{2},$$

which shows that the identity is true for $n + 1$. By induction, therefore, it is true for all natural numbers.   □

6.3   In the chessboard tiling problem, the grids that were pictured were 8×8, like a regular chessboard. But do the arguments depend on this? Does the result hold for any size square grid?

Let us consider first $n \times n$ square grids, with two opposite corners deleted.

Theorem. *An $n \times n$ square grid, with two opposite corners deleted, can never be tiled with 2×1 dominoes.*

Proof. If n is even, then let us place a chessboard pattern upon the square grid. Since n is even, every full row has the same number of light and dark squares, and so the undeleted $n \times n$ square has equal numbers of light and dark squares. Furthermore, because n is even, adjacent corner squares will have opposite colors, and consequently opposite corners will have the same color. So after deleting two opposite corners, we will have fewer squares of one color than of the other. But every 2×1 domino will cover one light square and one dark square, and so there can be no tiling.

Alternatively, if n is odd, then the full $n \times n$ board has n^2 many squares, an odd number, and so the deleted square also has an odd number of squares, which therefore obviously cannot be tiled by 2×1 dominoes.   □

Let us now generalize the result to rectangles, with opposite corners deleted.

Theorem. *One can tile a rectangular $n \times m$ array, with two opposite corners deleted, using 2×1 dominoes if and only if one of the numbers is odd and the other is even.*

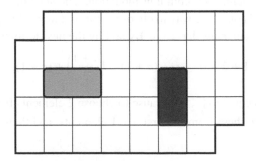

Proof. If both numbers are odd, then nm is odd and so $nm - 2$ also is odd. In this case, one cannot tile the figure with 2×1 dominoes, since each domino uses two squares and therefore the tiling would have to cover an even number of squares altogether. If both numbers are even, however, as here with the 10×4 board, then again consider placing chessboard pattern on the board. As in the argument above, the two opposite corner squares will have the same color, and so the number of light and dark squares on the deleted board will be unequal, preventing a tiling by 2×1 dominoes.

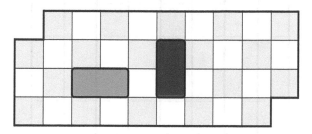

Last, consider the case with one side even and one odd, as with this 7×4 board.

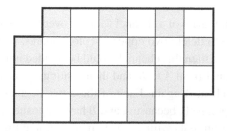

In this case, let us lay the board down so that there are an even number of rows, each with an odd number of squares, except for the top and bottom rows, which have an even number. We can tile the top and bottom rows with horizontal tiles. And there are an even number of rows between them, which can therefore be tiled with vertical dominoes. So we can tile the deleted board as desired.   □

6.4 Using the figure appearing in the chapter, prove that if you delete two opposite-color squares from an ordinary 8×8 chess board, you can tile it with 2×1 dominoes, no matter which two opposite-color squares are omitted. Does the argument generalize to any size square grid? What about rectangular grids?

Theorem. *If one deletes any two opposite-color squares from an 8×8 chessboard, the remaining board can be tiled with 2×1 dominoes.*

Proof. Consider the 8 × 8 chessboard, with black lines marked as indicated in the figure.

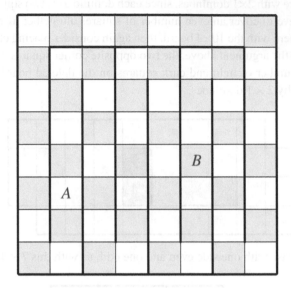

These lines depict the boundaries of a channel winding over the board, visiting every square exactly once. If one were to delete two opposite-color squares, such as at *A* and *B*, or any two opposite-color squares, then this channel would be divided into two segments. Imagine running through the channel from *A* to *B*, and then continuing further through the channel after *B* and finding oneself at *A* again. Each of these two segments starts with one color and ends with the other, precisely because *A* and *B* have opposite colors. Thus, we may tile each of the two channel segments with 2 × 1 dominoes, going first from *A* to *B*, and then skipping over *B* and going from *B* to *A*. So there is a tiling of the deleted board. And it doesn't matter which two squares are deleted, provided that they have opposite colors.   □

We can generalize this to arbitrary rectangular boards, provided that both sides have length at least 2 and at least one side has even length.

Theorem. *On any chessboard of size n×m, where these are positive integers, both at least 2, and at least one of them is even, then after deleting any two opposite-color squares, one may tile the remaining board with 2 × 1 dominoes. If both n and m are odd, this is never possible.*

Proof. The channel proof idea generalizes readily to larger boards. Place the even side horizontally, and define the channel as in the figure above by starting at the lower left, following a channel going up the left side, right across the top and down on the right side to the lower right corner, and then using the up-down zigzag pattern as in the 8 × 8 pattern above. (Note, if *m* = 2, then there is no actual zigzagging, and the channel simply returns along the bottom edge.) Since the horizontal side length is even, and each zigzag uses two columns, this zigzag pattern will return the channel to the lower left corner as desired. And then the proof described above shows that if one deletes any two opposite color squares, the remaining board can be tiled simply by staying within the channel.

Meanwhile, if both side lengths n and m are odd, then the board has an odd number of squares altogether, and so after deleting two squares, there are still an odd number of squares, and so it will be impossible to tile by 2×1 squares. □

Note also that an $n \times 1$ rectangle also does not have the desired property, if $n > 2$, since we can delete squares one away from each edge, and then what remains cannot be tiled with 2×1 dominoes.

> **6.5** Does the alternative 3-coloring of the chessboard, with one corner removed, still work with the argument that there is no tiling by 3×1 long dominoes?

The claim in the main text was that the one-square-deleted board at the left cannot be tiled with 3×1 dominoes. The proof offered there was the middle figure here. Each long domino will cover one square of each shade, but there are only twenty white squares, yet twenty-one medium-shaded squares and twenty-two dark-shaded squares, and so there can be no tiling.

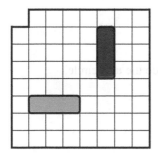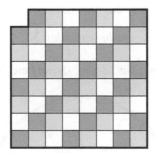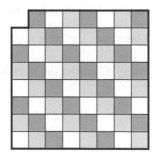

The question here is whether the argument still works with the figure at the right, and the answer is that this particular argument does not succeed with that particular shading pattern, because now there are exactly twenty-one squares of each shade. So it was important to use a shading pattern for which the argument succeeds.

6.7 Explain how the following figure can be used to prove that $2\pi > 6$ and hence that $\pi > 3$.

Theorem. $3 < \pi$.

Proof. The number π, by definition, is the ratio of the circumference of a circle to its diameter. Thus, in a circle of radius 1, the circumference is 2π. The perimeter of the inscribed hexagon, as in the figure, is 6, because the hexagon consists of six equilateral triangles, whose side lengths are the same as the radius of the circle. Since the edges of the hexagon are straight lines and the shortest way to travel from each of those six points to the next, it follows that the perimeter of the hexagon is smaller than the circumference of the circle. So $6 < 2\pi$, and consequently $3 < \pi$. □

Archimedes constructed approximations to π by extending the inscribed hexagon to an inscribed 12-gon, and then an inscribed 24-gon, and so on. The resulting perimeters converge to 2π from below. Let us carry this out one step.

Theorem. *Similarly,* $6\sqrt{2 - \sqrt{3}} < \pi$. *In particular,* $3.1 < \pi$.

Proof. Consider the inscribed regular 12-gon, shown here, inside the unit circle.

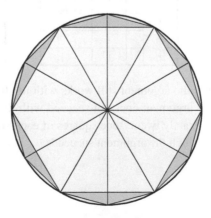

To form the inscribed 12-gon, each equilateral triangle of the inscribed hexagon is bisected by a radius of the circle, making twelve angles of 30° at the center. If you consider the small dark-shaded triangles that are added as a result, the leg bordering on the hexagon has length 1/2, since it is half of the original equilateral-triangle side; the other very small leg on the radius has length $1 - \sqrt{3}/2$, since it is what remains of the radius 1 after the equilateral-triangle altitude of $\sqrt{3}/2$. So by the Pythagorean theorem, the perimeter segment of the 12-gon, which is the hypotenuse of the small extra dark triangle, has length $\sqrt{(\frac{1}{2})^2 + (1 - \frac{\sqrt{3}}{2})^2}$, which some simple algebraic manipulation shows is equal to $\sqrt{2 - \sqrt{3}}$. Since there are twelve sides, and the perimeter of the circle is 2π, we may conclude that

$$6\sqrt{2 - \sqrt{3}} \quad < \quad \pi,$$

as desired. One may calculate the value of the square-root expression as $3.10582\ldots$, and so in particular, $3.1 < \pi$. $\qquad\square$

The inscribed 24-gon has a perimeter of approximately 3.132 and the inscribed 48-gon has a perimeter of approximately 3.139, and one thereby begins to see how to obtain increasingly better approximations to π this way.

> **6.9** Find a proof without words in another collection that particularly inspires you, and present it to the class. Make sure to provide sufficient explanation that the rest of the class can understand and appreciate it; it is likely that a superior explanation will arise if you actually do use some words.

The reader can readily find various collections of proofs without words. A good starting point is the collection of answers posted on MathOverflow at https://mathoverflow.net/q/8846. Let me offer the following argument of Tom Apostol. The figure here is a proof that there is no isosceles right triangle with integer sides. This implies, in particular, that $\sqrt{2}$ is irrational. Perhaps the proof is improved with a few well-chosen words? Very well, let me provide them. Suppose we are given an integer isosceles right triangle, as the large outer triangle here. Mark off the circle from one leg to the hypotenuse as

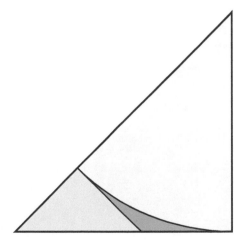

shown, and extend a perpendicular to form the smaller inner triangle, which is also isosce-les because it is a right triangle with a 45° angle. The two segments bordering the dark narrow region must have the same length, because they are both tangent to the circle. This length is the same as the other leg of the inner triangle, because it is isosceles. And so these are integers, since on the left it is cut from the original triangle's hypotenuse by a side, both of which have integer length. And also the hypotenuse of the inner triangle at bottom has integer length, being cut again from one integer segment by another. So we've produced a strictly smaller integer isosceles right triangle, and so there can be no smallest instance, a contradiction.

7 Theory of Games

The theory of games is mathematically rich, not only full of fascinating mathematical puzzles and conundrums but also having many deeper abstract results of enduring mathematical value. Yet, the subject is often strangely neglected in the traditional standard curriculum, despite an abundance of accessible and interesting problems paving the way to the more profound results. So let us explore the exercises on offer here, concerned with games introduced in the main text.

7.1 True or false: in the game Twenty-One, making a mistake early on doesn't matter so much, since you can recover by the end by adopting the winning strategy.

The statement is false. A single misstep in the game allows the opponent to take over in a way that ensures the opponent's success.

Theorem. *In the game of Twenty-One, any deviation from the strategy of ending with one of the numbers*

$$1 \quad 5 \quad 9 \quad 13 \quad 17 \quad 21$$

will result in a loss, if the opponent plays optimally after the deviation.

Proof. In the main text, we proved that it is a winning strategy to end one's play on those numbers. The opponent is unable to get from any of those numbers to the next number on the list in one move, but no matter what number the opponent does play from one of those numbers, one is thereby enabled oneself to get from *that* number to the next number on the list. Since the list includes the winning number 21, this strategy will result in a win.

If one should ever make a play in the game not ending on one of these numbers, then the opponent can immediately pick up the winning strategy of saying the numbers on the list. The reason is that any number below 21 that is not on the list is within 3 of the next number on the list, precisely because they are 4 apart. Therefore, any deviation from the winning strategy we described will result in a forced loss if the opponent should recognize the opportunity. $\square$

The same argument works in any of the modified games $G_{n,s}$ in the main text, where the players take turns saying up to the next s numbers, starting with 1, with the winner being whoever says n first. The winning strategy is to say the numbers with the same residue as n modulo $s + 1$, and any deviation from this will allow the opponent immediately to adopt the winning strategy. In this sense, the winning strategy in these games is unique.

7.3 Consider the game Fifteen, in which two players take turns selecting numbers from 1 to 15, each selected at most once; whoever first has three numbers adding to fifteen wins (it must be three numbers adding to fifteen, not two). Using the following matrix, prove that this game is isomorphic to tic-tac-toe. Conclude that there is no winning strategy.

$$\begin{pmatrix} 8 & 1 & 6 \\ 3 & 5 & 7 \\ 4 & 9 & 2 \end{pmatrix}$$

The American Museum of Mathematics in New York has a display where players can play this game—on one side, for the adults, you can see only the numbers $1, 2, 3, \ldots, 15$ listed linearly, but on the other side, for the kids, one sees the matrix, and the kids can usually win because of the tic-tac-toe insight.

I find it fun to observe players at MoMath, especially when quite mathematical parents, faced with the list of numbers rather than the tic-tac-toe layout, stand flummoxed, beaten again and again by their very young children. How is it possible? Perhaps, they think, their kids are geniuses? Indeed, the kids might very well be geniuses, but perhaps a more important reason, unknown at first to the parents, is that the children can see the tic-tac-toe layout and thus are more easily able to set up winning forking plays.

Theorem. *The game of Fifteen is isomorphic to tic-tac-toe, in the sense that each move in one of the games corresponds to a specific move in the other, in such a way that wins correspond to wins and draws to draws.*

Proof. Let us correspond moves in Fifteen to moves in tic-tac-toe using the matrix

$$\begin{pmatrix} 8 & 1 & 6 \\ 3 & 5 & 7 \\ 4 & 9 & 2 \end{pmatrix}.$$

That is, if we refer to the first player as X and the second as O, to make a move in Fifteen or in tic-tac-toe is to select a number or position in this matrix. In this way, every play of either game can be seen as a play in the other. Notice that this matrix is a magic square: every row, column, and diagonal adds to 15. Therefore, every win in tic-tac-toe will correspond to a win in Fifteen. Conversely, I claim, if three numbers do not lie on a row, column, or diagonal, then they do not add to 15. To see, notice that if one has two numbers in the same row or column, but the third is not, then the sum will not be 15 because the third number needs to be the other element of that row or column in order to add to 15. And if one has three numbers all in different rows and columns, then either one has a diagonal, or else one has one of the possibilities $8 + 9 + 7$, $3 + 1 + 2$, $3 + 9 + 6$, or $4 + 1 + 7$, none of which sum to 15. So the only way to win Fifteen is to have played numbers that correspond to a win in tic-tac-toe. Thus, the games are isomorphic by this correspondence. □

Since tic-tac-toe is well known not to have a winning strategy for either player, it follows that Fifteen also has no winning strategy.

7.5 Consider the modified version of the game Buckets of Fish, where on each turn a player may take a fish from one bucket and then either add or remove as many fish as desired from any of the buckets to the left. What is the winning strategy? Suppose that a player may add or take away at most one fish from each of the earlier buckets. What, then, is the winning strategy?

The difference is that in the original game one could only add fish to the buckets to the left, whereas in the modified game one can either add or remove fish from the earlier buckets.

Theorem. *In the modified game of Buckets of Fish, in which on each turn a player may take a fish from one bucket and then add or remove as many fish as desired from any of the buckets to the left, the winning strategy is to leave an even number of fish in each bucket. And the same is true for the further modified game in which one may add or take away at most one fish from each of the earlier buckets.*

Proof. First, we note that the proof of theorem 32 in the main text applies also to this modified game, showing that every play of the game is finite. The game will definitely end in finitely many moves, even though from many positions one can add fish to the buckets to the left.

Next, notice that if you are faced with a situation having an odd bucket, then indeed you can make all the buckets even, simply by removing one fish from the rightmost odd bucket and then either adding or removing fish to make all buckets to the left also even. So the all-even strategy is a strategy that can indeed be implemented and played on any position that is not already all even. Further, notice that if you give your opponent an all-even situation, with an even number of fish in each bucket, then he or she will necessarily cause an odd bucket on his or her next move, because the rightmost bucket that the player acts upon will have exactly one fish removed, thereby making it odd. So if you are ever able to give an all-even bucket situation to your opponent, then inductively you will be able to continue doing this until the game is over. And since the winning position, with no fish in any bucket, is an all-even position, it will be you who has won the game. Thus, the strategy to make all buckets even is a winning strategy in this game.

The same argument applies to the further modified game, where one may add or remove at most one fish from the earlier buckets, because you can make an odd number even by adding or subtracting one.   □

7.7 Consider a further modified version of the game Buckets of Fish, where on each turn a player may take any positive number of fish from any given bucket and add any number of fish to each of the buckets to the left. What is the winning strategy?

This difference, where one may now take any positive number of fish from a bucket, instead of just one, causes a fundamental change in the winning strategy and in which positions are winning. It will no longer be true that it is winning to give your opponent an all-even position, since the opponent could give you such an all-even position right back, by removing two fish from a bucket. So one needs to think carefully about what the new winning positions will be.

Theorem. *In the modified game Buckets of Fish, where on each turn a player may remove as many fish from a given bucket as desired and add any number of fish to each of the buckets to the left, the winning strategy is to leave the buckets with balanced adjacent pairs, so that the leftmost two buckets have the same number of fish, and also the next two, and the next two, and so on.*

Proof. Let us note that the proof of theorem 32 in the main text again shows that all plays of this modified game will end in finitely many moves. Next, suppose you are faced with a position that does not have balanced pairs in the sense described in the theorem statement. Then I claim that you can indeed make a move so as to create the balanced-pairs situation. Simply look for the rightmost violation of the balanced-pair requirement. One of the buckets in the pair has too many fish. So you may remove some fish from that bucket, so as to balance the pair, and then add fish, if necessary, to the earlier buckets in order to balance all the other pairs. So you may move on any unbalanced position so as to create the balanced-pair situation.

Observe next that if your opponent is given a balanced-pairs position, then whichever bucket the opponent removes fish from will become part of an unbalanced pair, since no other buckets have fish removed from them, and in particular, the partner bucket in the same pair as this bucket will no longer be balanced with it. It follows by induction that once one has played a balanced-pair position, one may continue to do so. And since the winning position, with all buckets empty, is a balanced-pair position, it follows that this is a winning strategy. □

7.9   What is the winning strategy for misère Nim? Prove your answer.

Recall that in Nim the winning player is the one who takes the last coin, while in misère Nim that is the losing move.

Theorem. *The winning strategy in misère Nim is the same as for Nim, except when such a move would lead to a position with all piles having only one coin, in which case the winning misère player should instead leave an odd number of piles.*

Proof. The strategy I am proposing for misère Nim is to leave balanced positions, except when all the piles have only one coin, in which case the misère player should leave an odd number of piles. This strategy is possible to implement, since any unbalanced position can be balanced, but if the balancing move would be to act on the very last multiple-coin pile, then the player can either leave one coin or none, so as to leave an odd number of single-coin piles.

Thus, the misère player can ensure always to give either a balanced position with at least one nontrivial pile (and hence two), or a position consisting of an odd number of single-coin piles. In particular, this implies that our misère player will never take the last coin, since the empty position does not have that form, and so this is a winning strategy.   □

How delightful and surprising that the misère Nim strategy is so similar to the ordinary-play Nim strategy! Perhaps one might have expected it to be totally opposite in some way—many people expect at first that the misère player should be playing unbalancing moves rather than balancing moves—but the theorem shows that no, the winning moves and positions in Nim and misère Nim are exactly the same until one gets down to the case of all trivial piles.

7.11   Give a complete analysis of the game of Chomp for a 3×2 chocolate bar. Would you rather go first or second? What is the winning play?

We know from theorem 59 in the main text that it must be the first player who has the winning strategy, and so we definitely want to go first. Part of the interest of that theorem, however, was that it was a pure-existence proof, showing that the first player can ensure a win, but it did not exhibit any specific uniform winning strategy. Here, in the 2×3 case, however, we can exhibit a specific winning strategy.

Theorem. *In the game of Chomp with a 3×2 chocolate bar, the first player can win. The strategy will be to chomp the lower-left square on the first move, and then either to match one white-dot chomp in the figure here for the other, or similarly with the shaded-dot chomps.*

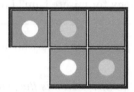

Proof. Let the first player chomp just the lower left square, as shown in the figure. The second player can respond with any of four chomps, other than the directly losing move of chomping the whole chocolate bar. If the second player chomps at one of the white-dot squares, then reply with the other, and similarly with the shaded-dot squares. The white-dot copying moves will reduce after this procedure to the shaded-dot squares position. And the shaded-dot copying procedure will lead inevitably to the opponent having to chomp the final corner square. So this is a winning strategy for the first player.   □

> **7.13** How large is the game tree for tic-tac-toe? Perhaps it is difficult to say exactly, but what are the best upper and lower bounds you can prove? Does the game have certain symmetries that allow you to understand the game with only part of the game tree?

Let's start slow and easy, and then work up to more precise bounds. We begin with a naive upper bound.

Theorem. *There are fewer than $9! = 362880$ many tic-tac-toe games.*

Proof. We obtain this naive upper bound simply by realizing that there are nine possible opening moves for X, and then eight replies by O, and then seven possible moves for X, and six replies, and so on. If we keep playing until the board is completely full, this would give $9! = 362880$ many possible games.   □

But of course, many tic-tac-toe games end before nine moves, and so this upper bound is much too large. Let us mount a more focused analysis.

Theorem. *The number of nodes on the first five levels of the tic-tac-toe game tree are exactly the numbers*

$$9 \quad 72 \quad 504 \quad 3024 \quad 15120.$$

For exactly 1440 *of the nodes on level 5, player X has just won the game.*

Proof. We simply observe that it will take at least three moves for X before a win is possible, and so the game tree is fully branching for the first five levels, giving 9 nodes after one move, $9 \cdot 8 = 72$ nodes after two moves, $9 \cdot 8 \cdot 7 = 504$ after three moves, and so on, until exactly $9 \cdot 8 \cdot 7 \cdot 6 \cdot 5 = 15120$ nodes on the tree after five moves. Some of these nodes are a win for X; let us calculate how many. Since there are 8 possible tic-tac-toes (3 possible rows, 3 possible columns, and 2 diagonals), and $\binom{6}{2} = 15$ many ways to place the Os outside those tic-tac-toes, this gives 120 possible boards on which X has won on the third play. Each such board can have been arrived at in 12 ways, since X might have played any of the three Xs on the first move, and then O either of the Os, and then X plays the second X, and then the rest is determined. So there are exactly $120 \cdot 12 = 1440$ many games in which X wins on his third move. $\square$

Theorem. *There are exactly* 54720 *many nodes on level* 6 *of the tic-tac-toe game tree (not counting the earlier wins for X). In exactly* 5328 *of them, player O has just won.*

Proof. According to our work above, there are precisely $15120 - 1440 = 13680$ many nodes on level 5 of the tree in which X has not yet won. Each of these positions has four possible moves for O, giving exactly 54720 nodes on level 6 of the tree (not counting the 1440 level-5 nodes where X has already won). Some of these nodes correspond to a win for O, but how many? There are as before 8 possible tic-tac-toes for O, each of which allows for $\binom{6}{3} = 20$ many choices for the Xs, or 160 boards with a tic-tac-toe for O. But some of those boards also show a tic-tac-toe for X, and so we shouldn't be counting them. How many? Well, each of the 6 row-or-column tic-tac-toes for O allows for two other possible tic-tac-toes for X (a diagonal tic-tac-toe for O rules out any tic-tac-toe for X). So this would be 12 boards after six moves showing a tic-tac-toe for each player (all of which would really be wins for X, since X would make it first). So there are exactly $160 - 12 = 148$ boards after six moves showing a tic-tac-toe for O but not for X. Each such board can be arrived at in $3 \cdot 3 \cdot 2 \cdot 2 = 36$ ways, giving exactly $148 \cdot 36 = 5328$ many nodes on level 6 that are wins for O. $\square$

Theorem. *There are exactly* 148176 *many nodes on level* 7 *of the game tree.*

Proof. According to our calculations above, there are $54720 - 5328 = 49392$ many nodes on level 6 for which neither player has yet won. Thus, there are exactly $49392 \cdot 3 = 148176$ many nodes on level 7 of the game tree. $\square$

Some of those positions are wins for O, and if we press on, we could calculate how many, but let us leave this here. I encourage those who are interested to press further. According to Wikipedia, the game tree for tic-tac-toe has exactly 255,168 many leaf nodes, each of which corresponds to a completed game.

Regarding symmetries, the game of tic-tac-toe has essentially only three opening moves: center, corner, side. To place an X in any corner square is isomorphic to placing it any other corner square, simply by rotating the board, since the winning configurations and the legal moves are not affected by rotation. Thus, for the purpose of analysis, one can immediately cut down the game tree by one-third. And then having placed the first X, in a corner, say, there will be up to symmetry only five possible replies, rather than eight, since by reflecting the board one can see that some moves come in pairs. Similarly, if X plays an edge move, there will be up to symmetry only five replies for O, and if X plays in the center, there are only two possible replies: edge or corner. In this way, one cuts down the tree considerably on the first few moves. But the full tree will still become quite large even when one using these symmetry ideas, since there are only eight symmetries of the board in all, and symmetries are generally lost after a few moves.

7.15 Criticize the play of both players in the tic-tac-toe game illustrated in figure 7.1 Did the winning player play well? On which move did the losing player go wrong?

The tic-tac-toe game in question is the following sequence of moves:

X		

		O
X		

X		O
X		

X		O
X		
O		

X		O
X	X	
O		

X		O
X	X	
O	O	

X		O
X	X	X
O		O

Player O made a mistake on his first move, because once the second X was played, player O had to reply to block the tic-tac-toe, and then the center X set up a forking double-win possibility, ensuring the win for X. Player O could have played in the center instead after the first X, and then it would have been possible to block to ensure a draw. Player X played perfectly after player O's opening blunder.

8 Pick's Theorem

Pick's theorem is a surprising pleasure, showing us how to compute the area of a polygonal region in the square lattice simply by counting the number of vertices on the boundary and in the interior of the figure. Here, we shall solve several exercises growing out of the theorem statement and its proof in the main text.

> **8.1** Give a direct proof, without using the results proved in this chapter, that Pick's theorem holds for triangles of height 1, whose base is parallel to one of the axes.

Theorem. *Every triangle of height 1 in the integer lattice, with base oriented to the axes, obeys Pick's theorem.*

Proof. Consider any such a triangle, with height 1 and a base of length n oriented to the axes.

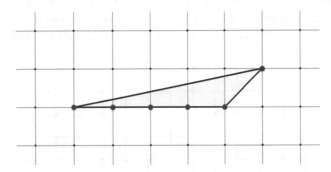

The area is $\frac{1}{2} \cdot$ base $\cdot$ height $= \frac{1}{2} \cdot n \cdot 1 = n/2$. There are $n + 1$ vertices on the base, and one more for the other vertex, making $n + 2$ many vertices on the boundary, and none in the interior. So $b = n + 2$ and $i = 0$, and Pick's formula therefore produces

$$i + b/2 - 1 = 0 + (n + 2)/2 - 1 = n/2,$$

which is indeed the area. So these triangles obey Pick's theorem. □

8.3 Can we omit the "simple" qualifier in Pick's theorem?

Theorem. *The identity of Pick's theorem does not hold for all polygonal figures in the integer lattice; it fails for some nonsimple polygonal figures.*

Recall that a *simple* polygonal figure in the integer lattice is one obtained from a closed-loop nonintersecting sequence of line segments joining lattice points. The statement of Pick's theorem in the main text applied only to simple polygonal figures.

Proof. Let us describe several kinds of counterexamples. First, for example, we might consider degenerate polygons. Consider a triangle with base length 5 and height 0. This is not a simple polygon because the line segment joining the beginning point to the end intersects the others.

This "triangle" has zero area, yet it has six boundary points (and no interior points), and therefore does not obey Pick's identity. Another degenerate counterexample is the empty polygon, with no vertices and no area. Since $A = b = i = 0$ for this figure, it will not obey $A = i + b/2 - 1$.

Perhaps more satisfying instances would be such figures as the bow tie or the annulus shown here. The bow tie is not a simple polygon, because the middle point is used twice and so the line segments are not nonintersecting. The annulus is not obtained from a single closed-loop sequence of segments but from two, one outside and one inside.

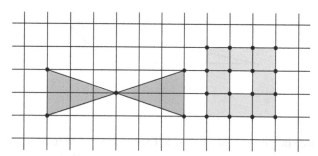

The bow-tie shape has area 6, yet $b = 7$ and $i = 4$, and so this does not obey Pick's formula. Similarly, the annulus has area 8, yet it seems to have $b = 16$ and $i = 0$, and so this also does not obey Pick's theorem.   □

8.5 Prove statement (2) of the key lemma 66.

Key Lemma. *(2) Suppose that P and Q are simple polygons in the integer lattice, which do not overlap but which are joined for a connected stretch of one or more edges on their boundaries. Let PQ denote the amalgamated polygon obtained by joining them together. If Pick's theorem holds for the amalgamation PQ and for one of P or Q, then it holds also for the other.*

Proof. Suppose that we join together two nonoverlapping simple polygons P and Q by fusing them along a common connected boundary, thereby forming the amalgamated polygon PQ.

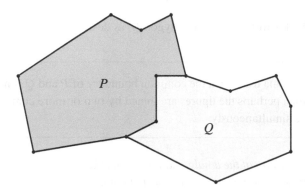

As in the main text, let us use the notation A_P, A_Q, i_P, i_Q, and b_P, b_Q to refer to the area, interior, and boundary lattice point counts of P and Q, respectively, and similarly with A_{PQ}, i_{PQ}, and b_{PQ} for the amalgamated polygon. We have assumed that Pick's theorem holds for PQ and for one of the polygons, let us say P. Thus, we have assumed that

$$A_{PQ} = i_{PQ} + \frac{b_{PQ}}{2} - 1$$

and

$$A_P = i_P + \frac{b_P}{2} - 1.$$

The area of the amalgamated polygon PQ is simply the sum of the areas of P and Q separately,

$$A_{PQ} = A_P + A_Q.$$

Let c be the number of boundary lattice points on the common boundary of P and Q, not including the two end points of the common boundary. It follows that

$$i_{PQ} = i_P + i_Q + c,$$

since the common boundary points (except for the two end points) become interior to the amalgamated region.

Similarly,

$$b_{PQ} = b_P + b_Q - 2c - 2,$$

since the $c + 2$ common boundary points were each counted twice, but only two of them remain on the boundary of the amalgamated region PQ.

Putting all of this together, we may calculate

$$A_Q = A_{PQ} - A_P$$
$$= (i_{PQ} + \frac{b_{PQ}}{2} - 1) - (i_P + \frac{b_P}{2} - 1)$$
$$= (i_{PQ} - i_P) + (b_{PQ} - b_P)/2$$
$$= (i_Q + c) + (b_Q - 2c - 2)/2$$
$$= i_Q + \frac{b_Q}{2} - 1.$$

And this verifies Pick's formula for the polygon Q, as desired. $\qquad\square$

8.7 Is the key lemma true when the common boundary of P and Q is not connected? For example, perhaps the figures are joined by two or more connected common boundaries simultaneously.

Theorem. *The conclusion of the amalgamation lemma (key lemma) is not necessarily true when the two regions are joined on a disconnected boundary.*

Proof. Let us join the two regions shown here. The boundary consists of two separate line segments, not connected.

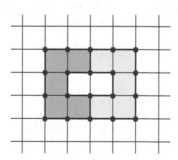

Each shaded region fulfills Pick's theorem on its own. But the annular amalgamated region does not, because it has area $A = 10$, with $b = 20$ boundary points and no interior points, $i = 0$. This does not fulfill $A = i + b/2 - 1$. $\qquad\square$

There is another sense, however, in which the amalgamation key lemma remains true even without assuming that the joining boundary is connected. Namely, if we assume that *P*, *Q*, and *PQ* are simple polygons in the lattice, then it follows from this that the amalgamation process must have been along a connected boundary, for otherwise the amalgamated shape would have a hole in it, preventing it from being a simple polygon.

8.9 Prove that every rectangle formed by vertices in the integer lattice (not necessarily oriented with the axes) has an area that is an integer.

In fact, we shall prove a stronger result.

Theorem. *Every parallelogram formed by vertices in the integer lattice has integer area.*

Proof. Consider any parallelogram in the integer lattice.

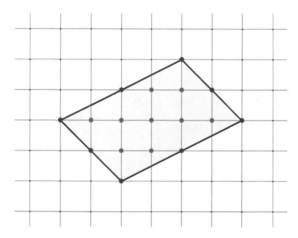

We may assume it is nondegenerate, for otherwise the area is zero, which is an integer. By Pick's theorem, the area is $i + b/2 - 1$, where i is the number of interior vertices and b is the number of lattice points on the boundary. The number i is an integer. And for a nondegenerate parallelogram, I claim, the number b is an even integer, because there are four points at the corners, and the other boundary points come in pairs, since any point on one side has a counterpart on the opposite side. So $b/2$ also is an integer, and so altogether the area is an integer. $\square$

There is another sense, however, in which the amalgamation law formula remains true even without assuming that the joining boundary is connected. Namely, if we assume that P, Q and PQ are simple polygons in the lattice, then it follows from this that the amalgamation process must have been along a connected boundary, for otherwise the amalgamated shape would have a whole in it, preventing it from being a simple polygon.

8.9 Prove that every rectangle formed by vertices of the integer lattice, not necessarily oriented with the axes, has an area that is an integer.

In fact we shall prove a stronger result:

Theorem. Every parallelogram spanned by vertices in the integer lattice has an integer area.

Proof. Consider any parallelogram in the integer lattice.

We may assume it is nondegenerate, for otherwise the area is zero, which is an integer. By Pick's theorem the area is $A = I + \frac{1}{2}B - 1$, where I is the number of interior lattice points and B is the number of boundary lattice points. Since the parallelogram has number I interior lattice points, and number B of the boundary lattice points, these points at the vertices and the boundary points occur in pairs, and so the area has a consequent integer value, since both I and $\frac{1}{2}B$ are integers. So the area is an integer.

9 Lattice-Point Polygons

In this chapter, we explore the nature of regular polygonal figures that might be found in various kinds of lattices, including the integer lattice, the triangular lattice, the hexagonal lattice, and other more general lattices.

> **9.1** Which regular polygons can be found using vertices of the 2 × 1 brick tiling? Prove your answer.
>
>

Theorem. *The figures that can be formed with lattices in the infinite 2 × 1 regular brick lattice are exactly the same as those in the 1 × 1 integer lattice, because these are in fact the same lattice. Consequently, the only regular polygons to be found in the brick lattice are squares.*

Proof. Consider the 2 × 1 regular brick tiling, where by *regular* I mean that on each subsequent row, the midpoint of each brick lines up with the edge between two bricks on the next row. If you consider carefully the lattice points of the brick lattice, the main observation to be made is that they are exactly the same as the lattice points of the 1 × 1 integer lattice.

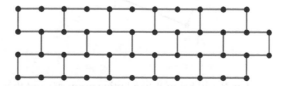

The point is that the long-side midpoints as well as the corners are lattice points in the brick lattice, because each such midpoint joins the corners of bricks in the next row. For the purpose of finding those vertex points, therefore, we are in effect tiling with the bricks cut in half, with 1×1 tiles, and this gives us the integer lattice. The figures to be found in the brick tiling, consequently, are the same as those in the integer lattice. For regular polygons, theorem 70 in the main text shows that only squares are to be found.   □

9.3   Prove that in the square lattice, any line segment joining two lattice points is the side of a square whose vertices are lattice points.

Theorem.   *In the square lattice, any line segment joining two lattice points is the side of a square whose vertices are lattice points.*

Proof.   We may consider the integer lattice, consisting of points (x, y) where x and y are integers. Suppose that A and B are lattice points. Notice that if we rotate the entire lattice by 90° around any lattice point, the lattice lands directly upon itself, with the rows landing on columns and columns on rows. Consider a counterclockwise such rotation around the point A. This rotation takes the point B to a point B'.

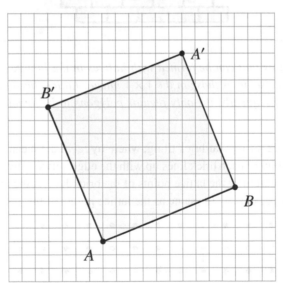

A similar rotation centered at B will bring A to a lattice point A', and these four points form a square, since the distances are the same and the angles are 90°. So AB is the side of a lattice-point square.   □

9.5 Prove that the square lattice is not necessarily invariant under the reflection swapping two lattice points.

It is clear that the reflection swapping any two vertices on the same horizontal row or the same vertical column will preserve the lattice. And also, the reflection swapping any two vertices on a 45° diagonal from each other will preserve the lattice. But it turns out that these reflections are the only lattice-preserving reflections.

Theorem. *The square lattice is not invariant under all the reflections swapping any two lattice points.*

Proof. It suffices to find such a reflection that doesn't take the lattice to itself. Consider the reflection swapping two opposite corners of a 2 × 1 rectangle. We have oriented the lattice here on an angle, so that the two points *A* and *B* lie horizontally, and so the reflection is through the dotted vertical line between them.

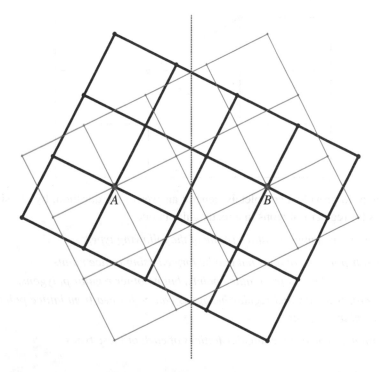

Thus, the original lattice is reflected to the new lattice, and clearly, the transformation does not take lattice points to lattice points. □

9.7 Can some nonsquare regular polygons arise from lattice points in some rectangular lattice, not necessarily square? Exactly which regular polygons can arise in such a lattice?

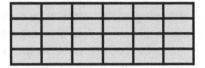

First, it is easy to see that some nonsquare regular polygons can be made in a rectangular lattice. For example, consider a unit rectangle of size $\sqrt{3} \times 1$. This is exactly the right distance to form an equilateral triangle as follows:

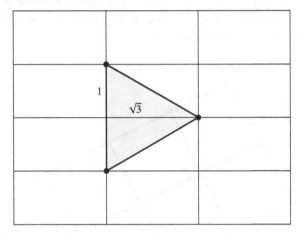

Let us attempt to provide a completely general answer to the question, by classifying the possibilities for regular polygons in a rectangular lattice.

Theorem. *Every rectangular lattice is one of the following types:*

1. *No regular polygons of any kind can be formed using lattice points.*
2. *Squares can be formed from lattice points, but no other regular polygons.*
3. *Equilateral triangles and regular hexagons can be formed from lattice points, but no other regular polygons.*

And furthermore, there are rectangular lattices of each of these types.

Proof. First of all, observe that by theorem 76 in the main text, we can never find any regular polygon in any kind of lattice, including any rectangular lattice, except possibly triangles, squares, and hexagons. So we need to consider only these three cases. Also, equilateral triangles and regular hexagons are to be found together, by the arguments in the main text, since from any equilateral triangle we may form a regular hexagon by translating along lattice line segments. And if we may form a hexagon, we may also form one twice as big (by side length), thereby forming a hexagon whose center is also a lattice point. But this will make six equilateral triangles. So the triangles and hexagons come together.

Next, let us argue that squares and equilateral triangles can never arise in the same rectangular lattice. Suppose that a rectangular lattice contains a square. Take two adjacent vertices of the square, not on the same vertical line. So their horizontal displacement d is an integer multiple of the base width w of the unit rectangle, $d = nw$. The adjacent edge of the square has exactly the same vertical displacement d, which must be an integer multiple of the base height h of the unit rectangle $d = mh$. So $nw = mh$ and therefore $w = \frac{m}{n} \cdot h$ are commensurable; both w and h are integer multiples of the common unit length h/n. So we can refine the rectangular lattice to a square lattice, using that common unit. Theorem 70 in the main text shows that there are no equilateral triangles to be found in that integer lattice, and so none can be found in this rectangular lattice either.

Last, let us show that all three types of rectangular lattices exist. First, the square lattice is a rectangular lattice having squares but no other regular polygons, as proved in the main text. Next, we observed above that in the rectangular lattice with unit rectangle of size $\sqrt{3} \times 1$, one may find equilateral triangles. The arguments above show that this lattice will therefore also exhibit regular hexagons, but not squares. Finally, consider a rectangular lattice using a base rectangle $1 \times h$, where h is irrational and not a rational multiple of $\sqrt{3}$. For example, let h be transcendental. In this case, there can be no square, since as previously we saw that if a rectangular lattice has a square at lattice points, then the base and height of the unit rectangle will be commensurable, which they are not here. If there is an equilateral triangle, then we may assume one vertex is at the origin. Let A be another vertex of the triangle not on the same vertical line as the first point. So it is at coordinate (n, mh) for some integers n and m. The third vertex B is obtained by rotating A by $60°$, and so using the rotation matrix we see that it has x-coordinate $\frac{1}{2}n - \frac{\sqrt{3}}{2}mh$. This must be an integer, in order for the point to be on the lattice, and so $\sqrt{3}mh$ also must be an integer k, which means $h = k/(m\sqrt{3}) = k\sqrt{3}/(3m)$, contrary to the choice of h. So for such a value of h, there can be no equilateral triangles and hence no regular hexagons, and also no squares. So all three types of rectangular lattices occur. □

9.8 Prove that in any rectangular lattice, using a rectangle whose side lengths are commensurable, the only regular polygon to be found using lattice points is a square.

Theorem. *The only regular polygon to be found at lattice points in a rectangular lattice, based on a unit rectangle having commensurable side lengths, is a square.*

Proof. Consider a rectangular lattice based on a unit rectangle having commensurable side lengths. Commensurability means that the sides have a common unit measure; each of them is an integer multiple of some common unit length, perhaps very small. Let us imagine constructing a square lattice using that common unit length. Because of commensurability, our rectangular lattice points will arise as points in that fine square lattice. In other words, the rectangular lattice is refined by the square lattice based on the common unit. And so any regular polygon to be found in the rectangular lattice can also be found in the square lattice. By theorem 70, the only such figures are squares.   □

9.9 Using colored chalk on the tiled plaza in the town's market square, a child connects the centers of some of the hexagons like this:

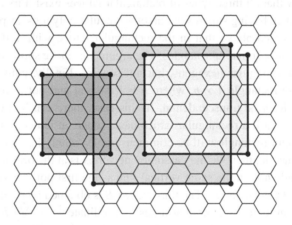

Has she made any squares?

Well, since this is a child playing with mathematical figures, I am likely to encourage her regardless and ask her also to draw other kinds of figures; the "squares" are surely square enough for a child's purpose. But if she is mathematically curious and serious about the question, wanting to know whether in fact any of them are *exactly* square, as a mathematical fact, then I should be delighted to discuss it further with her. The answer is that one cannot actually make a perfect square by joining the centers in a hexagonal lattice. In the image, the child's rectangles are aligned with the hexagonal sides, but even if she had drawn them with an inclined orientation, she still will not have made any squares.

Theorem. *No square can be made by joining the centers of hexagons in the hexagonal lattice.*

Proof. We refer here, of course, to nondegenerate squares. By including the center points of the hexagons into the lattice we would form exactly the triangular lattice, because each hexagon consists of six equilateral triangles formed by joining its center to its outer vertices. If we could form a square using the hexagonal centers in the hexagonal lattice, therefore, then we would be able to form a square in the triangular lattice. But theorem 73 shows that there is no square to be found in the triangular lattice. And so we cannot have found such a square using the centers of hexagons in the hexagonal lattice. □

Alternative proof. An alternative argument could proceed without the triangular lattice observation by observing that the centers of the hexagons are all uniformly offset from a hexagonal vertex by the same vector. So by translating the figure, we would find a square using lattice points in the hexagonal lattice, which we know is impossible by theorem 73. □

Meanwhile, let us also observe that by making the attempted square large enough or, equivalently, by using a fine enough hexagonal lattice, we can make a rectangle that is as close to being square as desired. For example, we can calculate the aspect ratios of the nearly square figures the child has already drawn. If the hexagons have unit side length, then each unit hexagon has width 2 and height $\sqrt{3}$. The small dark rectangle therefore has size $6 \times 4\sqrt{3}$, for an aspect ratio of 1.15. The medium-sized rectangle is $9 \times 5\sqrt{3}$, for a slightly better aspect ratio of 0.96. And the large rectangle is $12 \times 7\sqrt{3}$, for an aspect ratio of 1.01, which is very nearly square.

9.11 Prove that there are no regular polygons, except squares, in the rational plane $\mathbb{Q} \times \mathbb{Q}$.

Theorem. *No regular polygons, except squares, can be formed by vertices in the rational plane $\mathbb{Q} \times \mathbb{Q}$.*

Proof. Suppose a figure is formed by finitely many points in the rational plane $\mathbb{Q} \times \mathbb{Q}$. Since the rational numbers are closed under integer multiples, we may find a scaled-up instance of this figure by multiplying through by a common multiple of the denominators of the rational numbers arising as coordinates of points in the figure. Thus, we find a scaled-up version of the figure in the integer lattice. Since there are no regular polygons except squares in the integer lattice, it therefore follows that there can be none in the rational plane $\mathbb{Q} \times \mathbb{Q}$. □

9.13 Prove that the ratio of any two parallel line segments joining points in the integer lattice is a rational number. Is this true without the "parallel" qualifier?

Let us begin with the first part of the question.

Theorem. *The ratio of the lengths of any two parallel line segments joining points in the integer lattice is rational.*

Proof. Let us first give a soft proof. Consider a line segment in the integer lattice, such as the segment AB pictured here. The length of the line segment will be an integer multiple of the length of the most basic segment having that slope. In the instance here, for example, AB has slope $2/3$ and consists of three segments of the 3×2 diagonal.

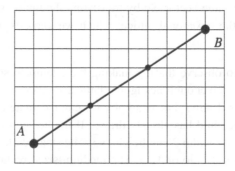

Since the lengths of any two parallel segments will thus be multiples of the same unit length in that direction, the ratio of their lengths will be rational. □

We can also give a messier but more explicit proof using coordinates.

Alternative explicit proof. Suppose that segment AB joins the point $A = (a_0, a_1)$ to the point $B = (b_0, b_1)$ and that it is parallel to segment CD joining $C = (c_0, c_1)$ to $D = (d_0, d_1)$. Since the segments are parallel, they have the common slope

$$\frac{b_1 - a_1}{b_0 - a_0} = \frac{d_1 - c_1}{d_0 - c_0}.$$

The respective lengths are

$$d_{AB} = \sqrt{(b_1 - a_1)^2 + (b_0 - a_0)^2}$$

and

$$d_{CD} = \sqrt{(d_1 - c_1)^2 + (d_0 - c_0)^2}.$$

Putting this together, we may compute the ratio of the lengths, and then factor out a term:

$$\frac{d_{AB}}{d_{CD}} = \sqrt{\frac{(b_1 - a_1)^2 + (b_0 - a_0)^2}{(d_1 - c_1)^2 + (d_0 - c_0)^2}}$$

$$= \sqrt{\frac{\left(\frac{b_1-a_1}{b_0-a_0}\right)^2 + 1}{\left(\frac{d_1-c_1}{d_0-c_0}\right)^2 + 1}} \left(\frac{b_0 - a_0}{d_0 - c_0}\right)$$

$$= \frac{b_0 - a_0}{d_0 - c_0}.$$

The compound radical expression in the middle drops out, because it is 1 by the common slope identity. Therefore, the length ratio is a rational number. $\qquad\square$

Observation. *The previous theorem is not true without the "parallel" qualifier.*

Proof. The ratio of the diagonal of the unit square to the side is $\sqrt{2}$, which is not rational. $\qquad\square$

> **9.15** Explain what happens precisely with the parallelogram argument proving theorem 76 when it is applied to triangles, squares, or hexagons.

Theorem 76 in the main text asserts that there are no regular polygons to be found in any lattice other than triangles, squares, and hexagons. The proof of this argument is concerned with regular polygons other than these three. One should form all the parallelograms from two adjacent edges of the regular polygon, taking in each case the resulting fourth point of the parallelogram, which we know must also be a lattice point; one thereby produces a smaller instance of the regular polygon. This is a contradiction, if one already had the smallest possible instance. And so for these kinds of regular polygons, there are none to be found in any lattice.

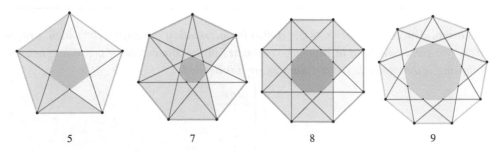

5   7   8   9

The argument breaks down with triangles, squares, and hexagons, however, as it must, since triangles, squares, and hexagons can be found in certain lattices. But let us see exactly how the argument breaks down; let us see what happens with the construction when it is applied to triangles, squares, and hexagons.

With an equilateral triangle, the vertices formed from the parallelograms constructed from adjacent edges will give you a much larger triangle, as indicated at the left. This breaks the minimality part of the argument, since the new triangle is not smaller than the original triangle, and so there is no contradiction.

With the square, the parallelograms formed by adjacent edges rebuild the very same square again, since the fourth vertex is already on the square. This also breaks the minimality argument, since the new square is not smaller than the original; it is indeed exactly the same as the original. With the hexagon, the fourth point of each parallelogram is the center point of the hexagon. Thus, the new vertices do not form another smaller hexagon at all, only a single point (a degenerate hexagon). So in each of the three cases, the argument toward contradiction breaks down, and no contradiction is made. This is of course to be expected, since we already know how these three kinds of regular polygons can occur in lattices.

9.17 For each integer $n \geq 3$, let $d(n)$ be the smallest dimension such that there is a regular planar n-gon realized at points of rational coordinates in dimension $d(n)$, if possible. For which values of n is $d(n)$ defined, and what is the complete list of values of the function $d(n)$?

Theorem. *For each integer $n \geq 3$, let $d(n)$ be the smallest dimension for which a regular n-gon can be realized at rational coordinates in that dimension, if possible. Then the following is the complete list of values where d is defined:*

n	$d(n)$
3	*3*
4	*2*
6	*3*

Proof. By theorem 76, we know that the only regular polygons to be found in any lattice whatsoever are triangles, squares, and hexagons, and so only these types of regular polygons will be found in the rational lattice of any dimension. So $d(n)$ is defined at most on $n = 3$, 4, and 6. Clearly, the square can be found in dimension 2, but not smaller, so $d(4) = 2$. Meanwhile, there are no triangles or hexagons in the two-dimensional integer lattice and therefore also not in the two-dimensional rational lattice. So $d(3)$ and $d(6)$ are at least 3. But in fact, they are both exactly 3, because one can form an equilateral triangle from three vertices of a cube, by slicing off one corner, and one can form a regular hexagon by joining the midpoints of certain sides in the cube.

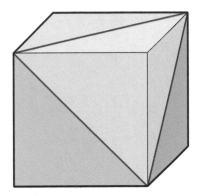 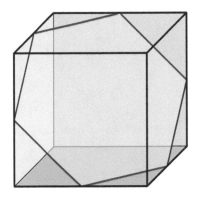

In fact, we have already argued in exercise 9.7 that once you have formed an equilateral triangle in a lattice, then you may also form a hexagon and conversely, and so we knew that $d(3) = d(6)$, and now we know furthermore that $d(3) = d(6) = 3$. □

10 Polygonal Dissection Congruence Theorem

The polygonal dissection theorem of Wallace, Bolyai, and Gerwein states that any two polygons of the same area are dissection congruent: One may cut one of them into finitely many pieces that can be rearranged to form the other. It is amazing! Let us solve several exercises exploring aspects of the proof of this theorem and its consequences and generalizations.

> **10.1** Explain what might go wrong in the proof of lemma 77.2 if we do not place the parallelogram on one of its longer sides.

Lemma 77.2 asserts that every parallelogram is dissection congruent to a rectangle. In the proof, we are instructed to place the parallelogram on one of its longest sides and make a vertical cut from one side to the other, thereby making two pieces that can be rearranged to form a rectangle as follows.

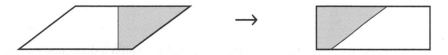

If one should place the parallelogram on a short side, however, then one will not necessarily find a vertical cut stretching from one side to the other.

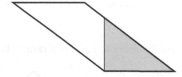

For this reason, therefore, the construction does not necessarily work with that modification. Since the figures still have equal area, however, one will still be able to achieve a dissection congruence. The point is that with the method of proof of lemma 77.2 described in the main text, one should simply place the parallelogram on one of its long sides, in order to achieve the desired congruence with one cut.

> **10.2** Prove the claim in the proof of lemma 77.3 that every rectangle is dissection congruent to a rectangle with no side more than twice as long as the other.

Theorem. *Every rectangle is dissection congruent to a rectangle with no side more than twice as long as the other.*

Proof. This theorem follows, of course, from the dissection congruence theorem (theorem 77), but it would be circular for us to appeal to that theorem here, since the point was that we used this easier theorem when proving the main dissection congruence theorem. So what is called for is a separate elementary proof of this result about rectangles.

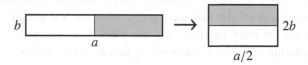

The main idea of the proof is that a long thin rectangle can be repeatedly cut in half and restacked to make a less extreme rectangle, and by repeating this, we shall achieve the desired rectangle. Specifically, suppose that we are given a rectangle of size $a \times b$, where $b \leq a$. If $a \leq 2b$, then we are already done, and so let us assume that $2b < a$, which means we have a long, thin rectangle. By slicing vertically in half, we may stack the two pieces and make a rectangle half as long and twice as high, of size $\frac{a}{2} \times 2b$. We keep doing this until for the first time the height of the rectangle is at least half the width. If we imagine having just achieved this (as pictured above), then we have $\frac{a}{4} \leq 2b$, and consequently both $2b < a$ and $\frac{a}{2} \leq 4b$, which shows that this $\frac{a}{2} \times 2b$ rectangle has both sides no more than twice the other. $\qquad \square$

> **10.3** What is the minimum number of pieces needed to exhibit a dissection congruence between a square and a 45-45-90 triangle?

Theorem. *Every square is dissection congruent to an isosceles right triangle using two pieces.*

Proof. Simply cut the square on the diagonal and rearrange the two pieces.

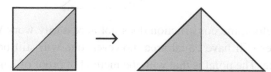

$\qquad \square$

And obviously, one piece does not suffice.

10.5 What is the minimum number of pieces needed to exhibit a dissection congruence between a square and two smaller squares?

First, let us provide an upper bound on the general case.

Theorem. *Any two squares are dissection congruent to a single large square using at most five pieces.*

Proof. The proof of lemma 77.4 shows how to dissect the two squares in order to form a single large square, and this dissection uses at most five pieces.

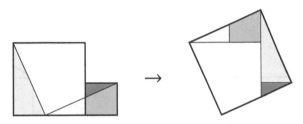

$\square$

Meanwhile, in special cases, one can get away with only four pieces, and this next theorem can be taken as the answer to the exercise.

Theorem. *Every square is dissection congruent to two smaller squares using at most four pieces.*

Proof. Cut the given square on the diagonals into four triangular pieces, which can be assembled into two smaller squares of the same size.

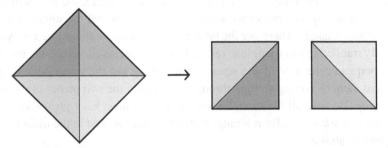

So this is an instance of dissection congruence between two small squares and a large square using four pieces. $\square$

Three pieces will never suffice, since in this case, one of the small squares will be left whole, and there is no way to place this square inside a larger square such that the remaining part can be cut in two pieces to form another square.

10.7 Using an ordinary piece of letter-size paper and scissors, carry out the algorithm
of the text to make it dissection congruent with a square.

Suppose we are given an $a \times b$ rectangle, where no side is more than twice the other, as
is true of ordinary letter-size or A4 paper. (If you have an exaggerated rectangle with one
side much longer than the other, then you can first apply the method of exercise 10.2 in
order to reduce to this case.) We aim to cut the rectangle into four pieces as shown at the
left and then rearrange them to form the square as at the right.

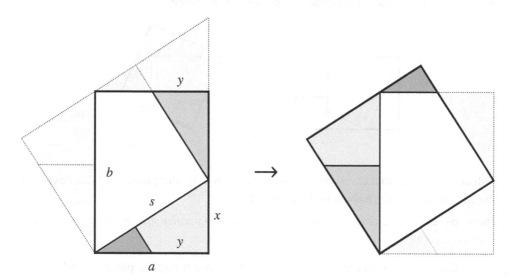

The hypotenuse s of the triangle at the bottom of the rectangle (at the left) will become a
side of the resulting square, and so we want $s^2 = ab$ in order that the square will have the
same area as the rectangle. Therefore the height x of that triangle should be $x = \sqrt{s^2 - a^2} =
\sqrt{ab - a^2}$. By translating this triangle to rest on top of the rectangle, as shown, one proceeds
to cut the perpendicular side of the square, first cutting out the upper triangle from the
rectangle and then, by cutting straight through, separating the two pieces of the translated
triangle on top. Because all the triangles are similar, we must have $y/(b - x) = x/a$, and
consequently $y = x(b - x)/a$. By rearranging the pieces, as proved in the main text, we can
form a square as shown.

With standard A4 paper, the rectangle is 210×297 mm, and so $x \approx 135$ mm and $y \approx 104$
mm, with the resulting square having sides $s \approx 250$ mm. With standard US letter-size
paper, the rectangle is $8\frac{1}{2} \times 11$ in, and so $x \approx 4.61$ in and $y \approx 3.47$ in, with $s \approx 9.67$ in. I
wonder whether there might be an elegant way to find these values not by measuring them
but by folding the paper in a certain way that would reveal the desired lines upon which to

cut. If you find an elegant geometrically correct folding-only construction solution to this problem, without any need for measuring, when using either A4 or letter-size paper, kindly send me an email describing it.

10.9 Use the figure of lemma 77.4 to give a dissection proof of the Pythagorean theorem.

Theorem (Pythagorean theorem). *In any right triangle, $a^2 + b^2 = c^2$, where the legs have lengths a and b and the hypotenuse has length c.*

Proof. Consider a right triangle with legs a and b and hypotenuse c, such as the labeled triangle in the figure here, used to prove lemma 77.4 in the main text. The two triangles are congruent, both having legs a and b and hypotenuse c. This figure shows how an $a \times a$ square and a $b \times b$ square together at the left are dissection congruent to a $c \times c$ square at the right.

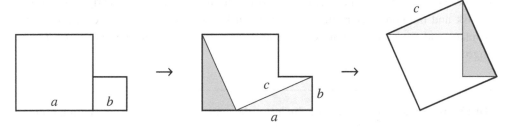

The rearranged shape is indeed a square with side length c, since all its sides arise as hypotenuses in the triangles and all the angles are right angles, being formed as they are from the acute angles of the original triangle. Since dissection is area preserving, it follows that $a^2 + b^2 = c^2$, as desired. □

10.11 Prove that no disk is scissors congruent to a noncircular ellipse.

Theorem. *No disk is scissors congruent to a noncircular ellipse.*

Proof. Suppose that we could cut a disk into finitely many pieces by cutting with scissors along smooth rectifiable curves and then rearrange them to form a disk. Consider the circular boundary of the disk. No fragment of this can be used to form part of the outer boundary of the ellipse, because no part of a noncircular ellipse is a circular arc; for example, the curvature of a noncircular ellipse is not constant on any part of the ellipse, but it is constant on a circular arc. So the circular boundary pieces from the disk must be used in the interior of the ellipse, where they will be matched with other pieces to fill out the interior of the ellipse. These other pieces must have in part concave circular boundaries, in order to fit together. In the ellipse, where the pieces are arranged, the total length of the convex circular-arc boundaries must therefore exactly match the total length of the concave circular-arc boundaries, because they are all used with each other matching up in the interior of the ellipse. But when the pieces were cut from the disk, each circular cut in the interior of the disk resulted in pieces that were convex on one side and concave on the other. And so there will be an excess of convex circular-arc boundaries coming from the boundary of the disk itself. So there can be no such way of cutting to make a disk scissors congruent to a noncircular ellipse.   □

Alternative proof using theorem 82.   The argument above basically reproves theorem 82, and so let us here instead simply appeal to the theorem instead of reproducing the proof of it. Theorem 82 asserts that if two figures are scissors congruent, then for every radius r, the difference between the convex and concave radius-r circular arc parts of their boundaries must be the same. If we take r as the radius of the disk, then this is clearly not true for the disk and the noncircular ellipse, since the disk has an excess of convex circular arc boundary and the ellipse has no circular arcs on its boundary. So they are not scissors congruent.   □

10.13 Prove that a sphere and a cylinder of the same volume are not dissection congruent (allow cutting along any plane).

Here, we allow you to cut the sphere along any plane, as though with a knife, like cutting cake, and then to rearrange the pieces. The question is whether by doing this with the sphere you can make pieces that can be rearranged to form a cylinder.

Theorem. *A sphere and a cylinder are never dissection congruent by the dissection process that allows cutting pieces along any plane.*

Proof.   Suppose that we could cut a sphere into finitely many pieces by slicing it with a plane, as though with a knife, and then rearrange them to form a cylinder. Since no part of the outer boundary of the cylinder is spherical, the pieces of the sphere having the outer spherical boundary must be matched when rearranged in the interior of the cylinder with other pieces, which must therefore have concave spherical surfaces. But with our cutting method, we cannot ever make such a piece, since all the cuts will have planar surfaces. So it is impossible.   □

Let us prove a stronger result by allowing more powerful cutting methods.

Theorem. *A sphere and a cylinder are not dissection congruent by a dissection process that allows cutting along any smooth surfaces.*

Proof. Suppose that we could cut a sphere into finitely many pieces, by cutting them along smooth surfaces, and then rearrange them to form a cylinder. Since no part of the outer boundary of the cylinder is spherical, the pieces arising from the sphere having part of the outer spherical boundary as a surface must be matched in the interior of the cylinder with other pieces, which must therefore have concave spherical surfaces. The total area of these surfaces must therefore be the same. In other words, the total area of the convex spherical surfaces of all of the pieces must be the same as the total area of the concave spherical surfaces of the pieces. And this must also be true for the pieces when they were cut from the sphere, but in the sphere there will be an excess of convex spherical surface arising from the original surface of the sphere. So it is impossible. □

> **10.15** Consider a generalized scissors congruence concept, allowing infinitely many pieces, provided that the total length of all cuts is finite. Does the analogue of theorem 80 still hold?

Theorem. *A square and a disk are not scissors congruent, even when infinitely many pieces are allowed, provided that the total length of all cuts is finite.*

Proof. The point here is that the invariant arguments used in the proof of theorems 80, 81, and 82 do not depend on the finiteness of the number of pieces but, rather, only on the finiteness of the total length of the cuts. If one could make cuts in a square of finite total length, then again the convex circular arc boundaries on the pieces must exactly match the concave circular arc boundaries, and so they cannot be rearranged to form a disk, which would similarly have to have an excess of convex circular arc boundaries. □

Meanwhile, if one doesn't insist that the total length of the cuts is finite, then the problem becomes subtle, and one must say more exactly what it means to have a piece and what it means to have a decomposition of a figure into pieces.

11 Functions and Relations

It is imperative for any aspiring mathematician to master the basic reasoning involved with functions and relations, to know whether one has an equivalence relation or not and to be able to prove it, and to know whether one has a function and, if so, whether it is injective or surjective or bijective. Thus, let us gain some practice proving things with these core mathematical topics.

> **11.1** Consider the collection of numerical expressions for rational numbers, like $\frac{3}{4}$ or $\frac{-6}{12}$. Let us consider these expressions not as numbers but as syntactic expressions $\frac{p}{q}$, pairs of integers, a numerator p and a nonzero denominator q, so that we count $\frac{1}{2}$ as a different expression than $\frac{2}{4}$. Define the relation $\frac{p}{q} \approx \frac{r}{s}$ for such expressions if they represent the same rational number, which happens precisely when $ps = rq$ in the integers. Prove that this is an equivalence relation.

Theorem. *The common equivalence of fractions of integers is an equivalence relation.*

Proof. We consider each integral fractional expression $\frac{p}{q}$ as a syntactic expression, with an integer numerator p and a nonzero integer denominator q. As fractions these are distinct, even if the rational number that they name turns out to be the same. The equivalence relation is defined by

$$\frac{p}{q} \approx \frac{r}{s} \quad \Longleftrightarrow \quad ps = rq.$$

In order to prove that this relation $\approx$ is an equivalence relation, we need to prove it is reflexive, symmetric, and transitive. The relation $\approx$ is reflexive because every fractional expression is equivalent to itself: $\frac{p}{q} \approx \frac{p}{q}$, since $pq = pq$ in the integers. Similarly, the relation $\approx$ is symmetric, since if $\frac{p}{q} \approx \frac{r}{s}$, then $ps = rq$, and consequently $rq = ps$ and therefore $\frac{r}{s} \approx \frac{p}{q}$, as desired. Finally, let us show that $\approx$ is transitive. Assume that $\frac{p}{q} \approx \frac{r}{s}$ and $\frac{r}{s} \approx \frac{t}{u}$. Therefore $ps = rq$ and $ru = ts$. Multiplying the first of these equations by u on both sides, and applying the second via substitution, we see that $psu = rqu = (ru)q = (ts)q = tsq$. The resulting identity is $psu = tsq$, and by canceling s, which is not zero, we deduce $pu = tq$, and so $\frac{p}{q} \approx \frac{t}{u}$, as desired. So this is an equivalence relation. $\square$

11.3 Show that if ~ and ≡ are two different equivalence relations on a set *A*, then the corresponding sets of equivalence classes are not the same.

Theorem. *Distinct equivalence relations on a set have distinct sets of equivalence classes. Indeed, two equivalence relations are identical if and only if their sets of equivalence classes are identical.*

Proof. If ~ and ≡ are identical, then of course they have the same sets of equivalence classes. The content of the theorem is the converse direction. Suppose that ~ and ≡ are not the same. So there is some instance in which one of the relations holds but not the other. Without loss of generality, by swapping the two relations if necessary, let us presume that $a \sim b$ for some particular elements a and b, yet $a \not\equiv b$. In this case, $b \in [a]_\sim$, yet $b \notin [a]_\equiv$, and so the equivalence class of a is not the same for the two relations. For each relation, the equivalence class of a is the only class containing a as an element, and so it follows that the respective sets of equivalence classes are not identical, as desired. $\square$

11.5 Is the squaring function x^2 well defined with respect to congruence modulo 5? How about exponentiation 2^x?

The question can be interpreted in two different ways. A naive way to interpret the question is whether the squaring operation x^2, when applied to equivalent integers, gives the same output. Well, if what we mean by the *same output* is that the value of x^2 as an integer is the same, then clearly the answer is negative. For example, 3 is equivalent to 103 modulo 5, but 3^2 is 9, while 103^2 is 10609, and these are not the same integer. So equivalent inputs did not give the same output, and in this naive sense the operation is not well defined with respect to that equivalence relation.

But there is another natural interpretation of the question, a somewhat more sophisticated version of it, which will give a positive result, and for this reason it is usually this interpretation that is intended in similar circumstances. Namely, since we are considering the integers modulo 5, we might ask whether the squaring operation on integers always takes equivalent inputs to *equivalent* outputs. That is, we consider both the input and the output modulo the equivalence relation. For this version of the question, it turns out that the squaring function is well-defined modulo 5.

Theorem. *The operation of squaring $x \mapsto x^2$ on the integers is well defined with respect to congruence modulo 5.*

Proof. We shall prove that squaring equivalent integers gives equivalent results. Suppose that $x \equiv y$ mod 5, which means that $y - x$ is a multiple of 5. So $y = x + 5k$ for some integer k. Therefore $y^2 = (x + 5k)^2 = x^2 + 10xk + 25k^2$, which can be expressed as $x^2 + 5(2xk + 5k^2)$. Thus, x^2 and y^2 differ by a multiple of 5, and so $x^2 \equiv y^2$ mod 5, as desired. In other words, squaring a number is well-defined modulo 5. □

Meanwhile, both versions of the question are negative for the exponentiation operation.

Observation. *The exponentiation operation on the integers $x \mapsto 2^x$ is not well defined with respect to congruence modulo 5.*

Proof. Notice that 1 is equivalent to 6 modulo 5, but 2^1 is not equivalent to 2^6, since $2^1 = 2$ and $2^6 = 64 \equiv 4$ mod 5. So the operation of exponentiation does not always map equivalent numbers to equivalent numbers, and so it is not well defined with respect to this equivalence relation. □

11.7 Show that the correspondence of equivalence relations with partitions and the correspondence of partitions with equivalence relations provided in the proofs of theorems 92 and 93, respectively, are inverses of each other. That is, if $\sim$ is an equivalence relation on a set X and P is the partition of X arising from $\sim$ in the proof of theorem 92, then the equivalence relation $\sim_P$ arising from P in the proof of theorem 93 is the same as the original relation $\sim$.

Theorem. *If we associate every equivalence relation $\sim$ on a set X with a corresponding partition $P_\sim$ on X as in theorem 92, and we associate every partition P on X with a corresponding equivalence relation $\sim_P$ on X as in theorem 93, then these associations are inverse, meaning that*

$$\sim_{P_\sim} = \sim \qquad and \qquad P_{\sim_P} = P.$$

I would recommend that the reader read this conclusion carefully, since the notation here is dense with information. The notation $\sim_{P_\sim}$ refers to the outcome of the process where one starts with an equivalence relation $\sim$, forms the associated partition $P_\sim$, and then forms the equivalence relation associated to that partition, with $\sim_{P_\sim}$ being the result. And similarly with $P_{\sim_P}$, we begin with a partition P, form the associated equivalence relation $\sim_P$, and then form the associated partition of that equivalence relation.

Proof. If $\sim$ is an equivalence relation on X, then the associated partition $P_\sim$ provided by theorem 92 in the main text is the set of equivalence classes for $\sim$, which that theorem shows is a partition of X. And if P is a partition of X, then the associated equivalence relation $\sim_P$ provided by theorem 93 is defined by $a \sim_P b$ if and only if there is some $A \in P$ with $a, b \in A$. That is, points are equivalent when they lie together in a piece of the partition.

Let us show that $\sim_{P_\sim} = \sim$ for any equivalence relation $\sim$. Points are equivalent $a \sim_{P_\sim} b$ for this doubly constructed relation if and only if they lie in the same piece of the partition $P_\sim$; but the pieces of this partition are precisely the equivalence classes of the original equivalence relation $\sim$, and it is a general fact about equivalence relations that two points a and b are in the same equivalence class if and only if they are equivalent. So $a \sim_{P_\sim} b$ if and only if $a \sim b$, and so these two equivalence relations are identical.

Next, let us show that $P_{\sim_P} = P$ for a partition P. The pieces of $P_{\sim_P}$ are the equivalence classes of $\sim_P$, and two points are $\sim_P$ equivalent if they lie in the same piece of P. So the equivalence classes of $\sim_P$ will be precisely the pieces of P, and thus $P_{\sim_P}$ is the same as P. $\square$

11.9 Can a relation be both symmetric and antisymmetric?

Yes, surprisingly, it can.

Theorem. *There is a relation on a set that is both symmetric and antisymmetric.*

Proof. Consider the identity relation $=$ on any set A. So the relation has instances $a = a$ for every $a \in A$, and no other instances. This is symmetric, since if $x = y$, then of course $y = x$ as well. Similarly, if $x = y$ and $y = x$, then of course $x = y$, and so it is antisymmetric. $\square$

11.11 How are the concepts of asymmetric, antisymmetric, and nonsymmetric related to one another? Which imply which others? Which of the eight conceivable combinations actually arise? Provide relations exhibiting each possible combination, and prove any provable implications.

Theorem. *Concerning the properties of asymmetry, antisymmetry, and nonsymmetry:*

1. *Every asymmetric relation is antisymmetric.*
2. *No other direct implications are provable. Specifically, for each row of the following table, there is a relation exhibiting that pattern of features.*

Asymmetric	Antisymmetric	Nonsymmetric
Y	Y	Y
Y	Y	N
N	Y	Y
N	Y	N
N	N	Y
N	N	N

3. *Meanwhile, for nonempty relations, every nonempty asymmetric relation is also non-symmetric.*

Proof. First, let us prove that every asymmetric relation is also antisymmetric. Assume that R is asymmetric, which means that every instance of the relation $a \mathrel{R} b$ implies that the converse instance $b \mathrel{R} a$ is *not* true. To show that R is antisymmetric, we want to show that if $x \mathrel{R} y$ and $y \mathrel{R} x$, then $x = y$. But by asymmetry, the situation in this hypothesis can never occur, since if $x \mathrel{R} y$, then $y \mathrel{R} x$ is not true. So the relation is vacuously antisymmetric. So we have proved statement (1).

For statement (2), the claim is that no other direct implications are possible between these three notions. The previous implication rules out two of the eight combinations, and to prove that no other implications are possible, we'll prove that all six remaining combinations of these features are possible. Let's proceed row by row in the table.

- *Asymmetric, antisymmetric,* and *nonsymmetric.* There are many relations like this. Consider the $<$ relation on the integers. This is asymmetric because $a < b \implies b \not< a$, and therefore it is also antisymmetric. And it is clearly nonsymmetric.
- *Asymmetric* and *antisymmetric* but not *nonsymmetric.* The empty relation—the always-false relation, which has no instances—is vacuously asymmetric and antisymmetric. But it is also vacuously symmetric and hence not nonsymmetric.
- Not *asymmetric* but *antisymmetric* and *nonsymmetric.* Consider the order $\leq$ on the integers. It is not asymmetric, since $2 \leq 2$, but it is antisymmetric, since $x \leq y \leq x \to x = y$, and it is nonsymmetric, since $2 \leq 5$ but not conversely.
- Not *asymmetric* but *antisymmetric*, yet not *nonsymmetric.* Consider the identity relation $=$ on any nonempty set. This is not asymmetric, since $x = x$ does not fail conversely, but it is trivially antisymmetric, and also symmetric, since $x = y \to y = x$, and hence not nonsymmetric.
- Not *asymmetric*, not *antisymmetric*, but *nonsymmetric.* Consider the at-least-as-tall-as relation on the set of all people, measured to the nearest millimeter, say. This is not asymmetric, since there are two people of the same height. It is not antisymmetric, since there are two different people of the same height, and so each is related to the other, but they are not equal. And it is nonsymmetric, since if one person is strictly taller than another, then the first person is at least as tall as the second, but not conversely.
- Not *asymmetric*, not *antisymmetric*, and not *nonsymmetric.* Consider the within-one-meter-of relation on the set of all people at a particular moment in time. This is not asymmetric, since every person is within one meter of themselves, and for these instances of the relation, the converse instance also holds, since it is the same instance. It is not antisymmetric, since two different people can be within one meter of each other, perhaps standing on the New York subway, and so they are related to each other in both directions, but not identical. And the relation is not nonsymmetric since it is symmetric: if x is within one meter of y, then also y is within one meter of x.

Finally, for statement (3), if a relation R is asymmetric and nonempty, so that it has an instance $a\ R\ b$, then the converse relation $b\ R\ a$ must fail by asymmetry, and so this relation is nonsymmetric. This explains why we needed to use the empty relation in the second example above. ☐

11.13 Give a relation on five elements, using as few relational instances as possible, whose reflexive-transitive closure is the full relation. How many relational edges did you use? Make a general claim and argument for a domain with n elements.

In fact, we can do it optimally even by taking only the transitive closure rather than the reflexive-transitive closure.

Theorem. *There is a relation on five elements, having exactly five relation instances, such that the transitive closure of it is the full relation.*

Proof. Let the relation form a cycle amongst the five elements, as here at the left. This relation has exactly five relation instances. The transitive closure is shown at the right.

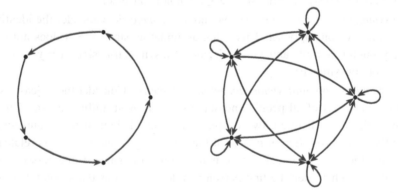

Clearly, from any node in the original cycle we can reach any node at all by following the cycle around. So the transitive closure of the cycle relation is the full relation, as pictured at the right. ☐

Since the transitive closure is already the full relation, the reflexive-transitive closure is as well.

Theorem. *No relation on a set of five elements with fewer than five relation instances can achieve the full relation as its reflexive-transitive closure.*

Proof. Suppose that R is a relation on a set of five points but has at most four relation instances. In this case, there must be at least one point a having no incoming relation instances, since otherwise we would see at least five relation instances. That is, there is no instance with $x \mathrel{R} a$. In this case, the transitive closure will not add any incoming relation instances to a, and so the reflexive-transitive closure of R will not have any incoming relation instances to a other than $a \mathrel{R} a$. So the reflexive-transitive closure will not be the full relation. □

Meanwhile, if we take the symmetric-transitive closure, we need only four relation instances to generate the full relation, since it would be enough to have a line from $a \to b \to c \to d \to e$, for then by symmetry we could also go in the other direction, and consequently every point can reach every other point through the symmetric-transitive closure.

11.15 What is the symmetric closure of the $<$ relation on the integers? What is the reflexive-symmetric closure of the $<$ relation on the integers?

Theorem. *The symmetric closure of the $<$ relation on the integers is precisely the unequal relation $\neq$.*

Proof. Every instance of the less-than relation $a < b$ is also an instance of the unequal relation $a \neq b$, and so $<$ is contained within $\neq$. Also, $\neq$ is a symmetric relation, since $a \neq b$ implies $b \neq a$. So the symmetric closure of $<$ is contained within $\neq$. But also, it must contain $\neq$, since if $a \neq b$ in the integers, then either $a < b$ or $b < a$, and so a will be related to b by the symmetric closure of $<$. So the symmetric closure of $<$ is exactly $\neq$. □

Theorem. *The reflexive-symmetric closure of $<$ on the integers is the full relation on the integers.*

Proof. Let $\leq$ be the reflexive-symmetric closure of $<$ on the integers. Since this includes all instances of the transitive closure of $<$, which is $\neq$ by the previous theorem, we know that if $a \neq b$, then $a \leq b$. But since we are taking the *reflexive*-transitive closure, we also have $a \leq a$ for every a. So in fact, any two integers are related by $\leq$, which is therefore the full relation on the integers. □

11.17 What is the symmetric transitive closure on the integers of the relation: x is either 15 more than or 10 less than y.

Theorem. *The symmetric transitive closure of the either-*15-*more-than-or-*10-*less-than relation on the integers is the relation* ≡₅ *of congruence modulo 5.*

Proof. Let us denote the original relation by ≃ and the symmetric transitive closure by ≡. Every integer x is ≃ related to both $x - 15$ and to $x + 10$. In other words, from any x, we can subtract 15 or add 10 in order to find a ≃-related element. In the symmetric-transitive closure ≡, we can do this repeatedly, in any direction. So x will be related to $x - 30$ and therefore also to x itself, by adding 10 three times. So ≡ is reflexive, and therefore it will be an equivalence relation. But also, notice that x will be related to $x - 15$ and therefore to $x + 5$, by adding 10 twice. So we can add or subtract 5 from a number as much as we like and stay ≡-equivalent. In particular, if two numbers x and y are congruent modulo 5, meaning $x ≡_5 y$, then they will be ≡-equivalent, since we can get from x to y by adding or subtracting 5 some number of times. So ≡ includes all instances of congruence modulo 5. But it cannot include any other relation instances, since the original relation ≃ is contained within congruence modulo 5 and congruence modulo 5 is symmetric and transitive. So ≡ is precisely the relation of congruence modulo 5.   □

11.19 Suppose that equivalence relation E refines equivalence relation F. How do the E-equivalence classes relate to the F-equivalence classes? State and prove a theorem about it.

Theorem. *One equivalence relation E refines another F if and only if every E-equivalence class is contained within some F-equivalence class.*

Proof. Recall that an equivalence relation E on a set A is said to *refine* another equivalence relation F on that same set if $a \ E \ b$ implies $a \ F \ b$ for all points $a, b \in A$. In this case, the E-equivalence class $[a]_E$ is a subset of the F-equivalence class $[a]_F$, since $b \in [a]_E$ means that $a \ E \ b$, which implies $a \ F \ b$, which implies $b \in [a]_F$. So if E refines F, then every E-equivalence class is contained in an F-equivalence class.

Conversely, suppose that every E-equivalence class is contained in an F-equivalence class. In order to show that E refines F, suppose that $a \ E \ b$. By assumption, $[a]_E$ is contained in some F-equivalence class $[x]_F$. Since a and b are E-equivalent, it follows that $a, b \in [a]_E$, and consequently $a, b \in [x]_F$. So both a and b are F-equivalent to some x, and consequently also to each other. So $a \ F \ b$, and we have therefore proved that E refines F, as desired.   □

11.21 Prove that the inverse of a bijection from A to B is a bijection from B to A.

Theorem. *The inverse of a bijection from A to B is a bijection from B to A.*

Proof. Suppose that $f : A \to B$ is a bijective function from A to B. The inverse function $f^{-1} : B \to A$ is defined by $f^{-1}(y) = x$, if $f(x) = y$. This is a function, since for each y in B there is exactly one value of x for which $f(x) = y$. There is at least one such x, precisely because f is surjective; and there is at most one such x, precisely because f is injective. So the inverse function f^{-1} is indeed a function from B to A. The inverse function f^{-1} is surjective onto A, because for any $x \in A$, we may consider $y = f(x)$ and observe consequently that $f^{-1}(y) = x$. So the range of f^{-1} is all of A and so it is surjective. Finally, the inverse function f^{-1} is injective, because if $f^{-1}(y) = x = f^{-1}(y')$, then by the definition of f^{-1} we must have $f(x) = y$ and $f(x) = y'$, and consequently $y = y'$. So the inverse function f^{-1} is a bijection from B to A. $\square$

12 Graph Theory

Graph theory is a delightful fusion of the abstract and the concrete. The principle objects of study, graphs, are fundamentally abstract, concerned with a generalized notion of connection between vertices or nodes. Often, these graphs arise as abstractions of a notion of adjacency in some system, perhaps locations in a network or connectivity in a circuit or in a network of computers, or perhaps reactivity of chemical compounds in a certain process, or compatibility in some other sense. And yet, while the graphs are abstract, they can often be easily represented with dots and lines on paper. So one can easily play with numerous examples and comprehend graph-theoretic arguments and constructions by how they are implemented with particular graphs. For this reason, I find graph theory to be an ideal forum in which to learn how to write proofs.

12.1 Prove that if a finite graph admits an Euler circuit, then for any particular edge of the graph, there is such an Euler circuit ending with exactly that edge.

Theorem. *If a finite graph admits an Euler circuit, then for any particular edge of the graph, there is an Euler circuit ending with that edge.*

Proof. Suppose that a graph G admits an Euler circuit. This Euler circuit follows a sequence of edges $e_1, e_2, \ldots, e_n$ through the graph, each edge traversed in a particular direction in such a way that the ending vertex of each edge is the starting vertex of the next (and the ending vertex of the last edge is the starting vertex of the first) and such that, furthermore, every edge in the graph is traversed exactly once. Suppose that e is one of the edges of the graph. If $e = e_n$ is the last edge on the given circuit, then we have already found the desired Euler circuit ending with e. So assume that $e = e_i$ for some $i < n$. Consider the circuit beginning with the next edge, continuing along the same circuit until the end, and then picking up the circuit from the beginning up to e_i:

$$e_{i+1}, \ldots, e_n, e_1, e_2, \ldots, e_i$$

This is an Euler circuit, because each edge ends at the start of the next edge, even when passing from e_n to e_1, and it uses every edge exactly once. And since it ends on e_i, we have thereby proved the theorem.   □

12.2 Prove that in any finite graph, the total degree, meaning the sum of the degrees of all the vertices, is always even. Conclude as a corollary that no finite graph can have an odd number of odd-degree vertices.

Theorem. *In any finite graph, the total degree is twice the number of edges. In particular, it is even.*

Proof. Consider any finite graph, with a set of vertices and a set of edges between them. The total degree is the sum of the degrees of each vertex. Each edge connects one vertex to another (or to the same vertex) and therefore makes a contribution of exactly 2 to the total degree. So the total degree is simply twice the number of edges. And so it is even.   □

Corollary. *No finite graph can have an odd number of odd-degree vertices.*

Proof. If a graph had an odd number of odd-degree vertices, then the total degree would be odd, since the sum of any number of even numbers and an odd number of odd numbers is odd. But the theorem shows that the total degree in a finite graph is always even, and so this cannot happen.   □

12.3 Prove that in any finite graph with exactly two vertices of odd degree, every Euler path must start at one of those odd-degree vertices and end at the other.

Theorem. *In any graph with an odd-degree vertex, every Euler path must start at one odd-degree vertex and end at another.*

Proof. Suppose that G is a graph with at least one odd-degree vertex, and suppose that we have an Euler path p. By theorem 97 in the main text, this cannot be an Euler circuit, so it must start and end at different vertices. Except for the starting and ending vertices of the path, every time the path enters a vertex via one edge it must depart on another. Therefore, all the intermediate vertices, those arising during the course of the Euler path, must have even degree. And similarly, the starting vertex will have an unmatched outgoing edge, and the ending point will have an unmatched incoming edge, so both of these vertices will have odd degree.   □

In particular, if there are exactly two odd-degree vertices, then every Euler path must start at one of them and end at the other. And indeed, by theorem 99 in the main text, if a finite graph is connected and has two odd-degree vertices, then there will be an Euler path. The theorem we just proved is helpful when trying to find them, because it tells us where we must start and where we must end.

12.5 Let K_n be the complete graph on n vertices. This is the graph with n vertices and an edge between any two distinct vertices. Show for nonzero n that K_n has an Euler circuit if and only if n is odd. For which n does K_n have an Euler path?

Theorem. *If n is a positive integer, then the complete graph K_n on n vertices has an Euler circuit if and only if n is odd.*

Proof. Assume that n is a positive integer. Theorem 97 in the main text shows that a connected finite graph has an Euler circuit if and only if every vertex has even degree. The graph K_n is certainly a finite connected graph. And since each vertex in K_n is connected to all the others, every vertex has degree $n - 1$, which will be even just in case n is odd. So K_n has an Euler circuit if and only if n is odd. □

Theorem. *For any natural number n, the complete graph K_n on n vertices has an Euler path if and only if n is odd or $n \leq 2$.*

Proof. Theorem 99 in the main text shows that a finite connected graph admits an Euler path if and only if there are at most two vertices of odd degree. We pointed out in the proof of the previous theorem that every vertex of K_n has degree $n - 1$. If n is odd, therefore, we have no vertices of odd degree, and therefore there is an Euler circuit, which is also an Euler path. If $n \leq 2$, then we will have at most two vertices of odd degree, and so there will be an Euler path. If $n > 3$ and even, then we will have n vertices of odd degree, which is too many for there to be an Euler path. So K_n has an Euler path if and only if n is odd or $n \leq 2$. □

Let me call attention to the fact that this second theorem includes the case $n = 0$, the empty graph, which has an Euler circuit if we include the empty circuit. Since 0 is even, this is why we restricted to positive integers in the first theorem above.

12.6 Suppose that in the city of Königsberg, the North Prince lives in a castle on the north bank and the South Prince lives in a castle on the south bank. The university is on the east island, but the center of the townspeople's mathematical discussions is, as we mentioned, at the pub on the center island, with challenges and late-night attempts to "walk the bridges." Now, the North Prince has a plan to build a new bridge, an eighth bridge, in such a way that from his castle he could traverse all eight bridges and end at the pub, for celebration. Where should he build a bridge that would enable him to do this? Prove that there is essentially only one location for such a bridge, in terms of its connectivity, and furthermore, if such a bridge were built, the South Prince would definitely not have the same ability from his own castle.

Theorem. *If the North Prince were to build a bridge from the university to the south bank, then he could walk the bridges from his home to the pub. And furthermore, such a bridge is the only way to achieve this feature with only one new bridge.*

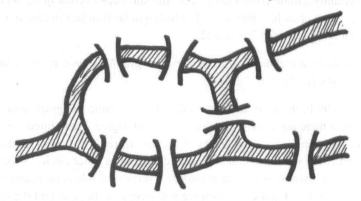

Proof. In the original Königsberg configuration, shown here, all the vertices have odd degree. If the North Prince were to build another bridge from the university (on the east island) to the south bank, parallel to the existing bridge at the lower right, then those two vertices would come to have even degree, making the north bank and the center island the only two vertices of odd degree. Thus, with such a new bridge, there would be an Euler path starting on the north bank and ending at the center island. In other words, the North Prince could walk the bridges from his home to the pub.

And indeed, if one were to achieve that feature with one new bridge, it would have to make the east island and the south bank have even degree, and so the new bridge must connect those two vertices. So such a bridge is the only way to do it. □

12.7 After the North Prince builds his bridge, the infuriated South Prince seeks revenge by building a ninth bridge, which will enable him to traverse all nine bridges exactly once, starting from his own castle and ending at the pub, while simultaneously preventing the analogous ability for the North Prince. Where should he build the ninth bridge?

Theorem. *After the North Prince erects a bridge from the university to the south bank, the South Prince could erect a bridge in the west across the river, from the south bank to the north bank, and thereby enable himself to walk the bridges from his home to the pub, while preventing the analogous ability for the North Prince.*

Proof. After the North Prince built the eighth bridge, connecting the university to the south bank, those two locations had even degree in the bridge graph, and the north bank and center island had odd degree. If the South Prince were to build a bridge from the south bank directly to the north bank, then only the south bank and center island would have odd degree. And so the South Prince would be able to walk the bridges on an Euler path from his home in the south to the center island. But there would be no Euler path from the north bank to the center island, because the north bank would have even degree. □

These exercises about additional bridges appear in the Wikipedia entry for the Königsberg bridge problem, which also relates the "walking the bridges" storyline.

12.9 Across the street from the five-room house considered in this chapter is another five-room house, pictured below, with a slightly different floor plan. Does this house admit a tour traversing each doorway exactly once? Does it admit such a tour starting and ending in the same place?

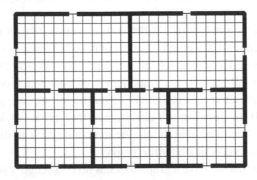

Theorem. *The house depicted above admits a tour traversing every doorway exactly once, starting outside and ending in the entrance foyer (the bottom center room). Indeed, every tour traversing every doorway once must start and end in those two locations, and in particular, there is no tour starting and ending in the same place.*

Proof. If we consider the associated graph, with one vertex for each room, including the outside location, which we can think of as one exterior "room," and with an edge between two vertices for each door that joins the corresponding rooms. If you inspect the graph, you will observe that every room has exactly four doors, except for the entrance foyer (bottom center), which has five doors, and the exterior room, which has nine. In the corresponding graph, therefore, there are exactly two vertices of odd degree, and so by the theorems in the main text, there is an Euler path starting outside and ending in the entrance foyer. Such an Euler path is a tour of the house in such a manner to traverse every doorway exactly once. Every such Euler path must start and end at those odd degree vertices, and so they must all either start outside and end in the entrance foyer or vice versa. □

12.11 Prove that if a graph has no odd-length cycles, then it is bipartite.

This exercise exhibits a common occurrence in mathematics, where one seeks at first to prove merely an implication $p \implies q$, but then realizes that the overlooked converse implication $q \implies p$ also holds—the two sides of the implication are equivalent, and we can prove the biconditional equivalence $p \iff q$. It is a sound mathematical habit of mind that whenever one has proved an implication, one should immediately ask, "Is the converse also true?" If it is, then one can establish a stronger mathematical connection between the statements by proving the biconditional equivalence; and if it is not, then this also would be good to know and prove.

Theorem. *A graph is bipartite if and only if it has no odd-length cycles.*

Proof. Recall that a graph is *bipartite* if the vertices can be partitioned into two disjoint sets, such that every edge of the graph, if any, connects a vertex in one set to a vertex in the other set. If a graph is bipartite, then let us imagine the two sets as the *red* vertices and the *blue* vertices. Thus, all edges in the graph join a red node to a blue node, and therefore every path in the graph alternates the colors of the nodes appearing along it, *red*, *blue*, *red*, *blue*, and so on. If the path forms a circuit, therefore, there would have to be an even number of nodes, in order to be able to join the final node back to the first; and therefore there are no odd-length cycles in a bipartite graph.

Conversely, suppose that we have a graph that has no odd-length cycles. Pick a vertex in each connected component of the graph (if the graph is connected, we will have picked a single vertex). Let us call these the *initial* vertices and color them red. Every vertex in the graph is connected by a path to exactly one initial vertex. Let us color every vertex red, if there is an even-length path to an initial vertex, and otherwise blue. Note that there cannot be both an even-length path from v to w and an odd-length path from v to w, for then we could traverse one of them from v to w and the other from w back to v, and this would be an odd-length cycle in the graph, which we have assumed does not happen. So every vertex gets colored red or blue, but never both. Furthermore, if a vertex v gets colored red, then its immediate neighbors will get colored blue, because the path from v to its chosen initial node will differ in length by exactly one from the path of its neighbors. Therefore, adjacent vertices will get opposite colors, and so the graph is bipartite. □

12.13 Formulate and prove a version of the Euler characteristic theorem that applies to finite planar graphs that are not necessarily connected. Your formula should involve $v - e + f$ and also refer to the number of connected components of the graph. Does your formula work with the empty graph?

Theorem. *Every finite planar graph obeys $v - e + f = n + 1$, where v is the number of vertices in the graph, e is the number of edges, f is the number of faces, and n is the number of connected components.*

Proof. Every finite planar graph G is the disjoint union of its connected components G_1, ..., G_n. Each component G_i is a connected finite planar graph and therefore obeys $v_i - e_i + f_i = 2$ by Euler's theorem. The total number of vertices v is the sum of the vertices v_i in each component, and similarly with the edges. All the interior faces of the graphs G_i are interior faces of G, but the single exterior face is counted as part of f_i for each connected component graph G_i, even though all those faces form the same exterior face of G. If we have n connected components, then we have counted the exterior face n times, but really we should have counted it only once. In other words,

$$v = v_1 + \cdots + v_n,$$
$$e = e_1 + \cdots + e_n,$$
$$f = f_1 + \cdots + f_n - (n - 1).$$

Putting this together, we see that

$$v - e + f = (v_1 + \cdots + v_n) - (e_1 + \cdots + e_n) + (f_1 + \cdots + f_n) - (n - 1)$$
$$= (v_1 - e_1 + f_1) + \cdots + (v_n - e_n + f_n) - (n - 1)$$
$$= 2 + \cdots + 2 - (n - 1)$$
$$= 2n - (n - 1)$$
$$= n + 1,$$

and this proves the theorem. □

Alternative direct proof. The previous proof derived the result as a consequence of the original Euler theorem. Let us now give a direct argument, essentially by reproving Euler's theorem, but in a more general context without assuming the graph is connected. The claim is that $v - e + f = n + 1$ for any finite planar graph. Let's prove this by induction on the number of edges. For the anchor, suppose we have a graph with v vertices and no edges. In this case, we have v vertices, $e = 0$ edges, $f = 1$ face, and $n = v$ connected components. This obeys the theorem, since $v - e + f = v - 0 + 1 = n + 1$, as desired. Suppose now that we have a graph G, and we add an edge to it. If this edge joins two previously disconnected components, then it will not add any faces, and we will have increased e by 1 and decreased n by 1, preserving the equation $v - e + f = n + 1$. The other possibility is that the new edge joins two already-connected vertices, in which case it will increase both e and f by 1, while preserving v and n, and this will therefore preserve $v - e + f = n + 1$. So the claim is true for all finite graphs by induction on the number of edges.   □

With this second proof, we can deduce the original Euler theorem that $v - e + f = 2$ for finite connected planar graphs as a corollary; we would not have been able to do this as a consequence of the first proof, since we had used the original Euler result in that proof.

Note that the theorem works fine with the empty graph, which has no vertices, no edges, a single face (the outside face), and no connected components. So $v = e = 0$, $f = 1$ and $n = 0$, which obeys $v - e + f = n + 1$, as desired. This calculation was part of the anchor case calculation of the second proof, which handled the $e = 0$ case, whether or not also $v = 0$.

12.15 Construct a triangulation of the torus, and compute the Euler characteristic of it.

For a triangulation of the torus, perhaps you might have something very rich in mind, such as the figure here.[1] This is indeed a triangulation of the torus, but with so many vertices, edges, and faces that it might be irritating to compute the Euler characteristic by hand with it. So let us consider instead a much simpler triangulation of the torus, with only one vertex and two faces. Consider the following square region, cut into two triangles by the diagonal. Imagine that you have cut out the square shape and that the material is very flexible and stretchy. We shall form it into a torus.

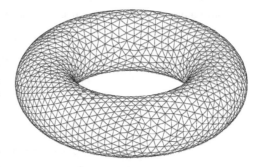

---

[1] Due to user Ag2gaeh, CC-BY-SA-3.0, available at https://commons.wikimedia.org/wiki/File:Torus-triang.png.

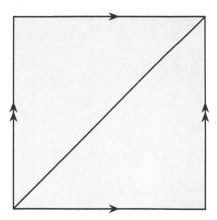

Roll the square from bottom to top, so as to bring the bottom edge into coincidence with the top edge, in effect identifying the two horizontal sides. This will make a small horizontal tube, with circular ends, corresponding to the double-marked edges of the square at the left and right. We may now bend this tube around so as to join the two circular boundaries, thereby making the familiar torus or doughnut shape. Ultimately, we have formed a torus from the square by identifying both the upper and lower edges and the left and right edges, respectively, with the indicated orientations. In particular, all four corners of the square have become identified with a single point on the torus. And so we have made a triangulation of the torus with just one vertex; we have three edges: the original diagonal edge, the top edge (same as the bottom edge), and the left edge (same as the right edge); and we have two faces, the two triangular regions visible in the square. Thus, $v = 1$, $e = 3$, and $f = 2$, which results in

$$v - e + f = 1 - 3 + 2 = 0.$$

The mathematical fact is that it doesn't matter how many vertices or triangles one uses, for in any triangulation of the torus, one will find $v - e + f = 0$.

The method of analyzing a topological space by means of a quotient topology, where one identifies various parts of a simpler space, such as the edges of a square as we have done, is widely used in the mathematical field known as algebraic topology.

13 Infinity

The allegory of Hilbert's Grand Hotel showcases the countability concept—a set is countable when it fits into Hilbert's hotel—and enables a gentle introduction to the abstract concepts of equinumerosity and cardinality. Here, we shall solve several exercises growing out of that allegory and the ideas it involves.

> **13.1** Suppose that guests arrived at Hilbert's hotel in the manner described in the chapter: first 6 guests arrive one by one, and then 1000 all at once, and then Hilbert's bus, followed by Hilbert's train, and finally Hilbert's half marathon. If the manager followed the procedure mentioned in the chapter, describe who are the occupants of rooms 0 through 100. How did they arrive and with which party? In which car or seat were they when they arrived, if they arrived by train or bus? Which fraction did they wear in the marathon? If you were the very first guest to arrive, where do you end up in the end? And what is the first room above you that is occupied? How did that guest arrive?

Let us assume that the guests arrive and are subsequently moved by the manager exactly as described in the main text. First 6 guests arrive one by one, and then 1000 all at once, and then Hilbert's bus, followed by Hilbert's train, and finally Hilbert's half marathon.

Let us trace the room trajectory of the original inhabitant of room n:

$$n \mapsto n+1 \mapsto n+2 \mapsto \cdots \mapsto n+6 \mapsto n+1006 \mapsto 2(n+1006) \mapsto 4(n+1006) \mapsto 8(n+1006)$$

Each guest was moved up successively by one room, six times in all, and then moved up by 1000 rooms, and then the room number was doubled (for Hilbert's bus), doubled again (for Hilbert's train), and finally doubled again (for Hilbert's half marathon).

Among the first six new guests, guest k, from $k = 0$ up to $k = 5$, follows the trajectory

$$0 \mapsto 1 \mapsto \cdots \mapsto 5 - k \mapsto 1005 - k \mapsto 2(1005 - k) \mapsto 4(1005 - k) \mapsto 8(1005 - k).$$

The *n*th guest amongst the 1000, from $n = 0$ up to $n = 999$, follows the trajectory

$$n \mapsto 2n \mapsto 4n \mapsto 8n.$$

The occupant of seat s on Hilbert's bus follows the room trajectory

$$2s + 1 \mapsto 2(2s + 1) \mapsto 4(2s + 1).$$

The occupant of car c, seat s on Hilbert's train follows the room trajectory

$$3^c \cdot 5^s \mapsto 2 \cdot 3^c \cdot 5^s.$$

And the runner p/q in Hilbert's half marathon finds him- or herself in room $3^p 5^q$.

Let us consider the first hundred rooms of the hotel just after the rooms have been occupied by the half-marathon runners. They are all put into the odd-numbered rooms, but not every odd-numbered room, since only those odd-numbered rooms with a number of the form $3^p 5^q$ are occupied by runners. Since the fractions are in lowest terms and $q \geq 1$, the only rooms like this up to room 100 are room $5 = 3^0 5^1$, occupied by runner $0 = 0/1$; room $15 = 3^1 5^1$, occupied by runner $1 = 1/1$; room $45 = 3^2 5^1$, occupied by runner $2 = 2/1$; and room $75 = 3^1 5^2$, occupied by runner $1/2$. Meanwhile, the even-numbered rooms are occupied by the guests who arrived with the group of 1000 or via Hilbert's bus or Hilbert's train. Guest n amongst the 1000 is in room $8n$. The occupant of bus seat s is now in room $4(2s + 1)$, which includes rooms 4, 12, 20, 28, 36, 44, 52, 60, 68, 76, 84, 92, and 100, for bus seats 0 through 12. The occupant of train car c, seat s now in room $2 \cdot 3^c \cdot 5^s$.

We can summarize the room occupation for rooms up to 100 in the following table:

Room	Room-number representation	Occupant
0	$8 \cdot 0$	guest 0 amongst the 1000
2	$2 \cdot 3^0 \cdot 5^0$	train passenger, car 0, seat 0
4	$4(2 \cdot 0 + 1)$	bus passenger, seat 0
5	$3^0 \cdot 5^1$	half-marathon runner $0 = 0/1$
6	$2 \cdot 3^1 \cdot 5^0$	train passenger, car 1, seat 0
8	$8 \cdot 1$	guest 1 amongst the 1000
10	$2 \cdot 3^0 \cdot 5^1$	train passenger, car 0, seat 1
12	$4(2 \cdot 1 + 1)$	bus passenger, seat 1
15	$3^1 \cdot 5^1$	half-marathon runner $1 = 1/1$
16	$8 \cdot 2$	guest 2 amongst the 1000
18	$2 \cdot 3^2 \cdot 5^0$	train passenger, car 2, seat 0
20	$4(2 \cdot 2 + 1)$	bus passenger, seat 2
24	$8 \cdot 3$	guest 3 amongst the 1000
28	$4(2 \cdot 3 + 1)$	bus passenger, seat 3
30	$2 \cdot 3^1 \cdot 5^1$	train passenger, car 1, seat 1
32	$8 \cdot 4$	guest 4 amongst the 1000
36	$4(2 \cdot 4 + 1)$	bus passenger, seat 4

40	$8 \cdot 5$	guest 5 amongst the 1000
44	$4(2 \cdot 5 + 1)$	bus passenger, seat 5
45	$3^2 \cdot 5^1$	half-marathon runner $2 = 2/1$
48	$8 \cdot 6$	guest 6 amongst the 1000
50	$2 \cdot 3^0 \cdot 5^2$	train passenger, car 0, seat 2
52	$4(2 \cdot 6 + 1)$	bus passenger, seat 6
54	$2 \cdot 3^3 \cdot 5^0$	train passenger, car 3, seat 0
56	$8 \cdot 7$	guest 7 amongst the 1000
60	$4(2 \cdot 7 + 1)$	bus passenger, seat 7
64	$8 \cdot 8$	guest 8 amongst the 1000
68	$4(2 \cdot 8 + 1)$	bus passenger, seat 8
72	$8 \cdot 9$	guest 9 amongst the 1000
75	$3^1 \cdot 5^2$	half-marathon runner $1/2$
76	$4(2 \cdot 9 + 1)$	bus passenger, seat 9
80	$8 \cdot 10$	guest 10 amongst the 1000
84	$4(2 \cdot 10 + 1)$	bus passenger, seat 10
88	$8 \cdot 11$	guest 11 amongst the 1000
90	$2 \cdot 3^2 \cdot 5^1$	train passenger, car 2, seat 1
92	$4(2 \cdot 11 + 1)$	bus passenger, seat 11
96	$8 \cdot 12$	guest 12 amongst the 1000
100	$4(2 \cdot 12 + 1)$	bus passenger, seat 12

All other rooms up to 100 are empty.

Since guest k among the first six guests ended up in room $8 \cdot (1005 - k)$, we can see that the very first guest ($k = 0$) ended up in room $8 \cdot 1005 = 8040$ at the end. The next guest ($k = 1$) landed in room 8048, and the intervening rooms are all empty, as none of them have a suitable form for their prime factorization: $8041 = 11 \cdot 17 \cdot 43$, $8042 = 2 \cdot 4021$, $8043 = 3 \cdot 7 \cdot 383$, $8044 = 2^2 \cdot 2011$, $8045 = 5 \cdot 1609$, $8046 = 2 \cdot 3^3 \cdot 149$, and $8047 = 13 \cdot 619$.

13.3 Consider a dozen or so of the most familiar functions on the real numbers seen in a typical calculus class. Which are functions from $\mathbb{R}$ to $\mathbb{R}$? Which are injective? surjective? bijective?

Let us consider in turn a dozen or so very common functions in a typical calculus class.

- The constant function with value $x \mapsto c$: We may view this is a function from $\mathbb{R}$ to $\mathbb{R}$, but it is not injective, because different input values of x yield the same output value c. It is not surjective, since its range has only one element c and therefore many other real numbers are omitted. Consequently, it is also not bijective.
- The identity function $x \mapsto x$: On domain $\mathbb{R}$, this is a function from $\mathbb{R}$ to $\mathbb{R}$, which is both injective and surjective, and hence also bijective.

- The squaring function $x \mapsto x^2$: This is often considered on domain $\mathbb{R}$, making it a function from $\mathbb{R}$ to $\mathbb{R}$, but it is not injective, because $(-3)^2 = 3^2$, and it is not surjective, because x^2 is never negative for real x. So it is not bijective.
- The cube function $x \mapsto x^3$: This makes an elegant function on the real numbers, both injective and surjective, and hence also bijective.
- The square-root function $x \mapsto \sqrt{x}$: In the real numbers, this function is defined only on the nonnegative real numbers, and so it is not a function from $\mathbb{R}$ to $\mathbb{R}$. On its domain it is injective, but it is not surjective onto $\mathbb{R}$, because $\sqrt{x}$ is taken to be the nonnegative square root, and so there are no negative numbers in the range.
- The absolute value function $x \mapsto |x|$: This is a function from $\mathbb{R}$ to $\mathbb{R}$, but it is not injective, since $|-2| = |2|$, and it is not surjective onto $\mathbb{R}$, because $|x|$ is never negative.
- The hyperbola $x \mapsto 1/x$: This function is not defined at 0, so it is a function not from $\mathbb{R}$ to $\mathbb{R}$ but from $\mathbb{R} \setminus \{0\}$ to $\mathbb{R}$. On that domain it is injective but not surjective to $\mathbb{R}$, since $1/x$ is never zero. But it is surjective to $\mathbb{R} \setminus \{0\}$, and thus it is a bijection of this set with itself.
- The sine function $x \mapsto \sin x$: This is a function from $\mathbb{R}$ to $\mathbb{R}$, which is periodic and hence not injective. It is not surjective, since the range is $[-1, 1]$.
- The cosine function $x \mapsto \cos x$: This function from $\mathbb{R}$ to $\mathbb{R}$ is a horizontal translation of the previous function and hence also is neither injective nor surjective.
- The tangent function $x \mapsto \tan x$: This function is not defined when $x = \pi k + \frac{\pi}{2}$, where the tangent function has a vertical asymptote. The tangent function is periodic and hence not injective, but even the principal branch, defined on the open interval $(-\pi/2, \pi/2)$, is surjective onto $\mathbb{R}$.
- The arctangent function $x \mapsto \tan^{-1} x$: One of the most graceful functions in calculus, this is the inverse function of the principal branch of the tangent function. That function is injective when restricted to domain $(-\pi/2, \pi/2)$, and surjective onto $\mathbb{R}$, and so the arctangent function is defined on all of $\mathbb{R}$. It is injective as a function from $\mathbb{R}$ to $\mathbb{R}$, but not surjective, since the range is exactly $(-\pi/2, \pi/2)$.
- The natural exponential function $x \mapsto e^x$: This function is defined on the whole real line, a function from $\mathbb{R}$ to $\mathbb{R}$, and it is injective but not surjective, since e^x is always positive for real numbers x.
- The natural logarithm function $x \mapsto \log x$, also commonly denoted $\ln x$ (the base 10 logarithm, hardly needed anymore in the computer era, would usually be denoted $\log_{10} x$ by mathematicians): This is the inverse of the exponential function e^x, defined on the range of that function, which consists of the positive real numbers. The log function is injective on its domain, and surjective onto $\mathbb{R}$, precisely because $\log(e^x) = x$.

> **13.5** Prove that if there is a surjective function $f : \mathbb{N} \to A$, then A is countable.

Recall that a set is defined to be countable when it is equinumerous with a set of natural numbers.

Theorem. *If there is a surjective function $f : \mathbb{N} \to A$, then A is countable.*

Proof. Suppose that $f : \mathbb{N} \to A$ is surjective. So every element of A appears in the list $f(0)$, $f(1)$, $f(2)$, and so on. Define $g : A \to \mathbb{N}$ by $g(a) = n$, when n is least for which $f(n) = a$. This is clearly injective, since different choices of a must get mapped to different n, and so A is bijective with a set of natural numbers—the range of g—and therefore is a countable set. □

A set A is countably infinite, meanwhile, if and only if there is a bijective function $f : \mathbb{N} \to A$.

> **13.6** Prove or refute: A set A is countable if and only if there is surjective function $f : \mathbb{N} \to A$. If this is false, state and prove a closely related true theorem.

This is one of those almost-true statements, which is basically right but not quite exactly right, because of an irritating exceptional case that sneaks in.

Claim. *The statement is false. There is a countable set A, with no surjection from $\mathbb{N}$ to A.*

Proof. This is a very picky point, but we can make a counterexample with the empty set $A = \varnothing$. The empty set is countable because it is bijective with a set of natural numbers (indeed, it already *is* a set of natural numbers—the empty set of numbers). But there is no function $f : \mathbb{N} \to \varnothing$ at all, since there is no possible value of $f(0)$. □

Meanwhile, we can prove the positive result by adding the relevant hypothesis.

Theorem. *A set A is countable and nonempty if and only if there is a surjective function $f : \mathbb{N} \to A$.*

Proof. Suppose that A is nonempty and countable. Since it is countable, there is a bijection between A and a set of natural numbers $g : A \to B \subseteq \mathbb{N}$. Pick a particular element $a \in A$, which is possible since A is nonempty. Let $f(n) = g^{-1}(n)$ if $n \in B$, and otherwise $f(n) = a$. Since g is bijective, every element of A has the form $g^{-1}(n)$ for some $n \in B$, and so f is a surjection from $\mathbb{N}$ to A.

Conversely, suppose that there is a surjection $f : \mathbb{N} \to A$. It follows that A is nonempty, since $f(17) \in A$, for example. For each $a \in A$, let $g(a)$ be the smallest number n with $f(n) = a$. Such an n exists since f is surjective. Different a will necessarily get different numbers n, and so g is injective. Thus, g is a bijection of A with its range, which is a set of natural numbers. So A is countable. $\square$

13.7   Prove that if A and B are sets, with B nonempty, and there is an injective function from B to A, then there is a surjective function from A to B.

Theorem. *If A and B are sets, with B nonempty, and there is an injective function from B to A, then there is a surjective function from A to B.*

Proof. Assume that A and B are sets, with B nonempty, and suppose that $f : B \to A$ is an injective function. Fix an element $b_0 \in B$, and define a function $g : A \to B$ according to the rule $g(a) = b$, if $f(b) = a$, and otherwise $g(a) = b_0$, if there is no such b. That is, g attempts to invert the function f, if the input a is indeed in the range of f, but otherwise sends a to the default value b_0. This is a function, since for any given a, if there is a b for which $f(b) = a$, then it is unique since f is injective. That is, the injectivity of f is precisely what we need for g to be functional. And g is surjective onto B, since for every $b \in B$ we have $g(f(b)) = b$. $\square$

13.9   Prove that the equinumerosity relation $\simeq$ is an equivalence relation.

Theorem. *Equinumerosity is an equivalence relation.*

Proof. The equinumerosity relation $\simeq$ is clearly reflexive, because every set A is equinumerous with itself, that is, $A \simeq A$, by means of the identity function $f : A \to A$, with $f(a) = a$. And the equinumerosity relation is symmetric, because if $A \simeq B$, then there is a bijective function $f : A \to B$; the inverse function $f^{-1} : B \to A$ is a bijection in the converse direction, and so $B \simeq A$, as desired. Finally, the equinumerosity relation is transitive, because if $A \simeq B$ via the bijection $f : A \to B$ and $B \simeq C$ via the bijection $g : B \to C$, then $A \simeq C$ via the composition function $g \circ f : A \to C$. The composition $g \circ f$ is injective, because the composition of injective functions is injective (theorem 94 in the main text), and it is surjective because the composition of surjective functions is surjective (exercise 11.20). So it is bijective, and therefore A is equinumerous with C as desired. Thus, the equinumerosity relation is reflexive, symmetric, and transitive, and so it is an equivalence relation. $\square$

> **13.11** Prove that if $A \leq B$ and B is countable, then so is A.

Theorem. *If $A \leq B$ and B is countable, then so is A.*

Proof. Assume that $A \leq B$, which means that there is an injective function $f : A \to B$, and that B is countable, which means that B is bijective with a set of natural numbers. Let $g : B \to C \subseteq \mathbb{N}$ be such a bijection. The composition function $g \circ f : A \to \mathbb{N}$ is injective, because it is the composition of injective functions (theorem 94 in the main text), and so A is bijective with the range of this composition function, which is a set of natural numbers. So A is countable, as desired. $\square$

> **13.13** Prove that $2^{\mathbb{N}}$ is equinumerous with the power set $P(\mathbb{N})$.

Theorem. *The Cantor space $2^{\mathbb{N}}$ is equinumerous with the power set $P(\mathbb{N})$.*

Proof. Recall that $2^{\mathbb{N}}$ is the set of infinite binary sequences. Every such binary sequence can be thought of as a function $s : \mathbb{N} \to \{0, 1\}$, where $s(n)$ is the nth binary bit of the sequence. For each such sequence s, let $A_s = \{n \mid s(n) = 1\}$, the set of places where s has digit 1. Another way to say this is that the sequence s is the *characteristic function* (also known as the *indicator function*) of the set A_s. Define $\pi : 2^{\mathbb{N}} \to P(\mathbb{N})$ by $\pi : s \mapsto A_s$. So we have associated to every sequence s in Cantor space the set A_s of which it is the characteristic function. This is a function from the Cantor space of binary sequences to the power set $P(\mathbb{N})$. The function is injective, since any two sequences differ on some nth binary bit, and so the associated sets will disagree on whether n is member. And the function is surjective, since every set $A \subseteq \mathbb{N}$ is A_s for the sequence s that is the characteristic function of A, defined by $s(n) = 1$, if $n \in A$, and 0 otherwise. So we have defined a bijection between $2^{\mathbb{N}}$ and $P(\mathbb{N})$, and so they are equinumerous. $\square$

> **13.15** Prove that the set of real numbers $\mathbb{R}$ is equinumerous with the set of complex numbers $\mathbb{C}$.

Theorem. *The set of real numbers $\mathbb{R}$ is equinumerous with the set of complex numbers $\mathbb{C}$.*

Proof. Theorem 118 in the main text shows that the real line $\mathbb{R}$ is equinumerous with the real plane $\mathbb{R} \times \mathbb{R}$. But the complex numbers are bijective with the plane, associating every complex number $a + bi$ with the point (a, b). So $\mathbb{R}$ is equinumerous with $\mathbb{C}$. $\square$

> **13.17** Prove that if A and B are equinumerous, then so are $A \times C$ and $B \times C$ for any set C. Use this to give an inductive proof of exercise 13.16.

Theorem. *If A is equinumerous with B, then $A \times C$ is equinumerous with $B \times C$ for any set C.*

Proof. Suppose that $f : A \to B$ is a bijection, witnessing that A and B are equinumerous. For any set C, define $g : A \times C \to B \times C$ by $g(a,c) = (f(a),c)$. That is, we apply the function f in the first coordinate and keep the second coordinate unchanged. This is injective, since different values of a will give different values of $f(a)$, and obviously changing c will change the output $g(a,c)$. And it is surjective, since if $(b,c) \in B \times C$, then since f is surjective we have $f(a) = b$ for some $a \in A$, and consequently, $g(a,c) = (f(a),c) = (b,c)$, as desired. So this is a bijection, and so $A \times C$ is equinumerous with $B \times C$. $\square$

We can now use this to give an inductive proof of the result of exercise 13.16.

Corollary. *The real line $\mathbb{R}$ is equinumerous with the plane $\mathbb{R} \times \mathbb{R}$ and with space $\mathbb{R}^3$, and indeed, with $\mathbb{R}^n$ in any finite dimension.*

Proof. We know already from theorem 118 in the main text that $\mathbb{R}$ is equinumerous with $\mathbb{R} \times \mathbb{R}$. By the previous theorem, it follows that $\mathbb{R} \times \mathbb{R}$ is equinumerous with $\mathbb{R} \times \mathbb{R} \times \mathbb{R}$, and so $\mathbb{R}$ is as well. More generally, if $\mathbb{R}$ is equinumerous with some $\mathbb{R}^n$, then $\mathbb{R} \times \mathbb{R}$ is equinumerous with $\mathbb{R}^n \times \mathbb{R}$, which is $\mathbb{R}^{n+1}$, and so $\mathbb{R}$ is equinumerous with $\mathbb{R}^{n+1}$. Thus, by induction, $\mathbb{R}$ is equinumerous with $\mathbb{R}^n$ in every finite dimension. $\square$

In fact, the real line $\mathbb{R}$ is equinumerous with the infinite-dimensional space $\mathbb{R}^{\mathbb{N}}$. One can see this by means of the following chain of equinumerosities:

$$\mathbb{R}^{\mathbb{N}} \simeq (2^{\mathbb{N}})^{\mathbb{N}} \simeq 2^{\mathbb{N} \times \mathbb{N}} \simeq 2^{\mathbb{N}} \simeq \mathbb{R}.$$

The first equinumerosity holds because $\mathbb{R}$ is bijective with Cantor space $2^{\mathbb{N}}$; the second holds essentially by a parenthesis-changing argument, showing that a function from $\mathbb{N}$ to $2^{\mathbb{N}}$ is essentially the same as a function from $\mathbb{N} \times \mathbb{N}$ to 2; the third holds because $\mathbb{N} \times \mathbb{N}$ is equinumerous with $\mathbb{N}$; and the last is again the equinumerosity of Cantor space with $\mathbb{R}$. Meanwhile, if one increases the dimension to $\mathbb{R}$, then $\mathbb{R}^{\mathbb{R}}$ is strictly larger than $\mathbb{R}$, since it is at least as large as $2^{\mathbb{R}}$, which is equinumerous with the power set $P(\mathbb{R})$, which is strictly larger than $\mathbb{R}$.

14 Order Theory

A certain mathematical maturity arises when one begins to treat the concept of order seriously. One distinguishes between partial orders and linear orders, between strict orders and reflexive orders, and this is not even to mention preorders. In this chapter, we solve several exercises that aim to develop this mathematical maturity.

14.1 Prove that if a partial order has two or more minimal elements, then it has no least element.

Theorem. *If a partial order has at least two minimal elements, then it has no least element.*

Proof. Suppose that a and b are distinct minimal elements of a partial order $\langle P, \leq \rangle$, and that c is a least element. Since c is least, it follows that $c \leq a$, since it is below everything in P. Since a is minimal, this implies $c = a$, for otherwise c would be strictly below a. By exactly similar reasoning, we can see that $c \leq b$ and consequently $c = b$ as well. This contradicts our assumption that a and b were distinct. $\qquad\square$

14.3 Provide a partial order with a unique maximal element but no largest element.

Theorem. *There is a partial order with a unique maximal element but no largest element.*

Proof. Consider the partial order consisting of the integer line $\mathbb{Z}$ under the usual order, together with an extra point z placed to the side, incomparable with all other elements. This order has a unique maximal element, the new element z, because there is nothing above it, whereas no other elements in the order are maximal, because every integer is below its successor. Meanwhile, this order has no largest element, because there is no element in the order that is above every integer. $\qquad\square$

It may be interesting to notice also that the element z used in the proof is also minimal in that order, because there is nothing strictly below it. So this is a partial order with an element that is both maximal and minimal yet neither smallest nor largest.

14.5 Prove the remarks made in the chapter about partial preorders, namely, that there are two different partial preorders $\leq_1$, $\leq_2$ with the same corresponding strict order $<_1 = <_2$. Conclude that with preorders, the $\leq$ relation has more information.

Theorem. *There are two different partial preorders $\leq_1$ and $\leq_2$ with the same underlying set and the same corresponding strict orders $<_1 = <_2$. Thus, in general, the strict order of a partial preorder does not determine the preorder.*

Proof. Let $A = \{a, b\}$ be a set with two elements, and consider the partial preorder $\leq_1$ by which they are incomparable, with neither element less than nor greater than the other, although both are $\leq_1$ related to themselves. In this order, there are no instances of the strict order, and so $<_1$ is the empty relation. Now consider the preorder $\leq_2$ by which $a \leq_2 b$ and $b \leq_2 a$, as well as being related to themselves. In this order, again, there are no instances of the strict order, and so $<_2$ is the empty relation. So we have two partial preorders on the set, whose strict orders are identical. $\square$

In general, the strict order $<$ of a preorder is not able to distinguish the case where two distinct elements are *equivalent*, in the sense that $a \leq b \leq a$, from the case where the two elements are *incomparable*, with $a \nleq b \nleq a$. This is precisely why $<$ can have less information than $\leq$ with preorders, precisely when there are such clusters of equivalent nodes. With partial orders, in contrast, as opposed to preorders, the first kind of case does not occur, since the antisymmetric property exactly rules them out, and with partial orders one can reconstruct $\leq$ on a given set uniquely from the information contained in $<$.

14.7 Prove that the isomorphism relation on partial orders is an equivalence relation.

Theorem. *The isomorphism relation on partial orders is an equivalence relation.*

Proof. In order to show that a relation is an equivalence relation, we must show that it is reflexive, symmetric, and transitive.

(Reflexive) Every partial order $\langle P, \leq \rangle$ is isomorphic to itself by the identity function $f(x) = x$, since in this case we clearly have $x \leq y \iff f(x) \leq f(y)$, and so the isomorphism relation is reflexive.

(Symmetric) If a partial order $\langle P, \leq_P \rangle$ is isomorphic to $\langle Q, \leq_Q \rangle$ by the isomorphism $f :$ $P \to Q$, then the inverse map $f^{-1} : Q \to P$ is an isomorphism in the converse direction, showing that $\langle Q, \leq_Q \rangle$ is isomorphic to $\langle P, \leq_P \rangle$, because the biconditional $x \leq_P y \iff$ $f(x) \leq_Q f(y)$ amounts exactly to $f^{-1}(a) \leq_P f^{-1}(b) \iff a \leq_Q b$, if we take $a = f(x)$ and $b = f(y)$. So the isomorphism relation is symmetric.

(Transitive) Finally, to see that it is transitive, suppose that $\langle P, \leq_P \rangle$ is isomorphic to $\langle Q, \leq_Q \rangle$ via $f : P \to Q$ and that $\langle Q, \leq_Q \rangle$ is isomorphic to $\langle R, \leq_R \rangle$ via $g : Q \to R$. It follows that $\langle P, \leq_P \rangle$ is isomorphic to $\langle R, \leq_R \rangle$ via the composition $g \circ f : P \to R$, which is a bijection, since the composition of bijections is bijective, and structure preserving, because $x \leq_P y \iff f(x) \leq_Q f(y) \iff g(f(x)) \leq_R g(f(x))$. So the isomorphism relation is also transitive, and thus it is an equivalence relation. $\square$

In fact, the theorem has little to do with partial orders, since an analogous theorem can be proved for any kind of mathematical structure. Namely, the isomorphism relation for any kind of mathematical structure is an equivalence relation.

14.8 Suppose that $\leq$ is a partial order. Define that elements p and q are *comparable* if either $p \leq q$ or $q \leq p$; otherwise, they are *incomparable*. Prove or refute: Comparability is an equivalence relation. Does your answer change if you know that the order is linear?

Theorem. *The comparability relation in a partial order is not necessarily an equivalence relation. While it is always reflexive and symmetric, there are partial orders whose comparability relation is not transitive.*

Proof. In any partial order, every element is comparable to itself, and if a and b are comparable, then so are b and a. So comparability is both reflexive and symmetric.

But it needn't necessarily be transitive. To see this, consider the partial order specified by the following Hasse diagram:

Notice that a is comparable to b, and b is comparable to c, but a and c are not comparable. This is a violation of transitivity for the comparability relation in this order. So comparability in this order is not an equivalence relation. $\square$

But notice that all elements in a linear order are comparable (indeed, that is the definition of linearity), and so the comparability relation in a linear order is the complete (always true) relation, which is an equivalence relation.

14.9 Can we omit the linearity assumption in theorem 125?

Theorem 125 asserts that every infinite linear order admits an infinite increasing sequence or an infinite decreasing sequence. We cannot omit the linearity assumption, in light of the following theorem.

Theorem. *There is an infinite partial order with no infinite increasing sequence and no infinite decreasing sequence.*

Proof. Consider the partial order consisting entirely of a single infinite antichain, with no nontrivial instances of the order relation. Thus, we place an order on an infinite set A by stating that $x \leq y$ only when $x = y$. This is a reflexive, transitive, and antisymmetric, and so it is a partial order. But it has no instances of strict inequality $x < y$, and so it has no increasing or decreasing sequences at all. □

14.11 Prove that the collection of all positive integers $\langle \mathbb{Z}^+, \mid \rangle$ under divisibility is not isomorphic to $\langle P_{\text{fin}}(\mathbb{N}), \subseteq \rangle$.

Theorem. *The collection of positive integers $\langle \mathbb{Z}^+, \mid \rangle$ under the divisibility relation is not isomorphic to the collection of finite sets of natural numbers $\langle P_{\text{fin}}(\mathbb{N}), \subseteq \rangle$ under the subset relation.*

Proof. These are both partial orders, since divisibility of positive integers and the subset relation are both reflexive, transitive, and antisymmetric. In addition, both orders have least elements, since 1 divides all positive integers and $\emptyset$ is a subset of every set. To show that the orders are not isomorphic, we shall find an order-structural property exhibited by $\mathbb{Z}^+$ but not by $P_{\text{fin}}(\mathbb{N})$; since isomorphic orders share all their order-structural properties, this will imply that the orders are not isomorphic. Notice that the number 4 has exactly three divisors: 1, 2, and 4. So the order $\langle \mathbb{Z}^+, \mid \rangle$ has an element with exactly three predecessors by the relation $\mid$. But $P_{\text{fin}}(\mathbb{N})$ has no such element with exactly three predecessors by the relation $\subseteq$, since a set of size n has 2^n many subsets, and this is never 3. So the two orders cannot be isomorphic, since any isomorphism f would have to map the predecessors of an element a in $\langle \mathbb{Z}^+, \mid \rangle$ exactly to the predecessors of the target element $f(a)$ in $\langle P_{\text{fin}}(\mathbb{N}), \subseteq \rangle$, while mapping nonpredecessors of a to nonpredecessors of $f(a)$, and therefore it would preserve the number of predecessors. So there can be no such isomorphism. □

14.13 Consider the collection of all intervals in the real numbers, including open intervals (a, b), closed intervals $[a, b]$, half-open intervals, and rays (a, ∞), $(-\infty, b]$, of all possible open/closed type and bounded/unbounded. Classify all these intervals up to isomorphism. How many different isomorphism types of orders do we find here?

Let us define that a set of real numbers is *convex* if, whenever it contains two real numbers, then it also contains all numbers between them. Every generalized interval, as described above, is convex, and indeed, these generalized intervals are exactly the convex sets of real numbers. So this problem is concerned with the convex sets of real numbers.

Theorem. *Every convex set of real numbers has exactly one of the following forms:*

1. *The empty set*
2. *A single point*
3. *A closed interval $[a, b]$, where $a < b$*
4. *An open interval (a, b), with $a < b$, including the intervals $(-\infty, \infty)$, $(-\infty, b)$, and (a, ∞)*
5. *A half-open interval of the form $[a, b)$, with $a < b$, including the case $[a, \infty)$*
6. *A half-open interval of the form $(a, b]$, where $a < b$, including the case $(-\infty, b]$*

Furthermore, all instances within each of these six cases are order isomorphic as linear orders, inheriting the order from the real numbers, and sets from distinct cases are not isomorphic. Thus, there are exactly six different order-isomorphism types of convex sets in the real numbers.

Proof. Suppose that A is a convex set of real numbers. If it has more than one point, then it will contain all points between any two of its elements, and so it will be an interval stretching from its infimum to its supremum. The particular case A fits into will be determined by whether it is unbounded above or below and by whether it has a largest or smallest element. What remains is to prove that all such intervals from the same case are isomorphic and that those from different cases are not isomorphic. There is only one empty set, of course, and so they are all isomorphic. And any two one-point linear orders are clearly isomorphic.

Consider now the case of closed interval orders. This is the case where the set A contains both its infimum and its supremum. We can linearly scale any nontrivial closed interval $[a, b]$ in the real numbers to any other such closed interval $[c, d]$, using the affine map $f(x) = c + (x - a)(d - c)/(b - a)$, which is order preserving and hence an isomorphism. So they are all isomorphic.

Similarly, all bounded open intervals (a, b) are isomorphic by means of the same linear scaling idea. For the unbounded open intervals, observe that the arctan function provides an order isomorphism from $(-\infty, \infty)$ to $(\pi/2, \pi/2)$, as well as from $(0, \infty)$ to $(0, \pi/2)$ and from $(-\infty, 0)$ to $(-\pi/2, 0)$. So all the various nonempty open intervals are order isomorphic.

The half-open intervals of the form $[a, b)$ are isomorphic by the same isomorphisms used above, except that we should include the left end point on both sides of the isomorphism. (Note that we do not allow $a = -\infty$ here, since we are concerned only with sets of real numbers.) The dual argument works with $(-\infty, b]$, showing that all half-open intervals of these forms are order isomorphic.

Finally, intervals chosen of different types from the six possibilities are not isomorphic, because any two intervals from different cases differ either in the question of whether they have 0, 1, or infinitely many elements, whether they have a maximal element, or whether they have a minimal element. Since these properties are preserved by isomorphism, the six cases are distinct up to isomorphism. And so there are exactly six isomorphism types of convex sets of real numbers. □

Note that we can achieve the empty set as an interval, using (a, b) when $b \le a$, for example, and we can achieve the one-point sets as closed intervals $[a, a] = \{a\}$. So the six cases are exactly the order-isomorphism possibilities for generalized intervals in the real numbers.

14.15 What is the analogue of theorem 130 for the strict eventual domination order $f <^* g$, for functions $f, g : \mathbb{N} \to \mathbb{N}$?

We can easily generalize Hausdorff's theorem to the strict eventual domination order. Define $f <^* g$ to hold for $f, g : \mathbb{N} \to \mathbb{N}$, if there is some number N such that $f(n) < g(n)$ for all $n \ge N$. That is, $f(n)$ is eventually strictly less than $g(n)$. Notice that this is a stronger condition than requiring merely that $f \le^* g$ and $f \ne g$, since these conditions could be true when f and g differed but still had infinitely many points of agreement $f(n) = g(n)$, a situation that would prevent $f <^* g$ from holding.

Theorem. *Every countable set of functions from $\mathbb{N}$ to $\mathbb{N}$ is strictly bounded in the strict eventual domination order. That is, given a countable list of functions f_0, f_1, f_2, ... on the natural numbers, there is a function $f : \mathbb{N} \to \mathbb{N}$ with $f_n <^* f$ for every n.*

Proof. We can follow essentially the same argument as the proof of Hausdorff's theorem (theorem 130) in the main text. Namely, let $f(n) = \sup_{m \le n} f_m(n) + 1$. Thus, f is a function form $\mathbb{N}$ to $\mathbb{N}$ and $f(n)$ is strictly bigger than $f_m(n)$ for all $m \le n$. In other words, for any fixed m, we have $f_m(n) < f(n)$ for all n beyond m. Thus, $f_m <^* f$ in the eventual domination order, and we have found our desired upper bound. □

We could alternatively have proved the theorem simply as a consequence of theorem 130 in the main text. Namely, that theorem provides a function $f : \mathbb{N} \to \mathbb{N}$ with $f_n \le^* f$ for every f. We may simply consider the function $f^+(n) = f(n) + 1$, which is strictly larger than f, and consequently $f_n <^* f^+$, as desired.

15 Real Analysis

Real analysis is the study of functions on the real numbers, particularly with respect to their continuity and differentiability features. Traditional developments of real analysis found the theory upon the least-upper-bound property, which expresses the fundamental completeness of the real number line. Here, we solve several exercises growing out of issues introduced in the main text, including the equivalence of the least-upper-bound property with the lesser-known principle of continuous induction, which we shall use to provide alternative proofs of several fundamental results.

> **15.1** Prove that the identity function and every constant function on the real numbers is continuous. Using theorems 132 and 133, prove by induction on the degree that every polynomial function is continuous.

Theorem. *Every polynomial function on the real numbers is continuous.*

Proof. Let us prove the theorem by induction on the degree of the polynomial. For the anchor case, consider a degree zero polynomial $f(x) = C$, a function with constant real value C. Let us prove carefully that this function is continuous. Fix any $\epsilon > 0$, and let $\delta = 17$; we could actually use any positive δ at all. If x and y are within δ, then $f(x)$ and $f(y)$ are both C, and so in particular, they are within ϵ of each other. So the function is continuous at any given point y. So every constant function on the real numbers is continuous.

Let us also prove that the identity function $g(x) = x$ is continuous. For any $\epsilon > 0$, we may take $\delta = \epsilon$ and notice that if x and y are within δ, then $g(x)$ and $g(y)$, which are just x and y again, of course, are within ϵ. So this function is continuous, as desired.

For the induction step, we shall use the fact that theorems 132 and 133 show that the sum and product of two continuous functions are continuous. Assume by induction that every polynomial of degree n is continuous, and consider a polynomial $g(x)$ of degree $n+1$. Thus, $g(x)$ has the form $a_0 + a_1 x + \cdots + a_{n+1} x^{n+1}$. Let us write this in the form

$$g(x) = a_0 + a_1 x + \cdots + a_{n+1} x^{n+1} = a_0 + x(a_1 + \cdots + a_{n+1} x^n).$$

Thus, we have realized $g(x)$ as the sum of a constant function a_0 with the product of the identity function x with a polynomial $a_1 + \cdots + a_{n+1}x^n$ of degree n. Since all three of these latter functions are continuous, and this is preserved by products and sums, it follows that g also is continuous. So by induction, every polynomial function on the real numbers is continuous. ☐

15.3 Find a function $g : \mathbb{R} \to \mathbb{R}$ on the real numbers that is discontinuous at every point except two, the points a and b.

Theorem. *For any two real numbers $a < b$, the function f defined here is discontinuous at every point except a and b.*

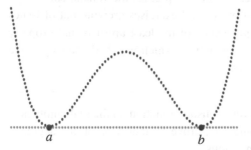

$$f(x) = \begin{cases} (x-a)^2(x-b)^2 & x \in \mathbb{Q} \\ 0 & x \notin \mathbb{Q} \end{cases}$$

Proof. The function is discontinuous at every point c other than a and b, because in every open neighborhood of c, there will be rational numbers x, where the function has value $(x-a)^2(x-b)^2$, and irrational numbers x, where the value is 0. Specifically, if $c \neq a, b$, it follows that $(c-a)^2(c-b)^2 \neq 0$, and so nearby rational values of x will have $f(x)$ very close to and sometimes above $(c-a)^2(c-b)^2$. But the irrational values of x will have $f(x) = 0$, and this is not within ϵ of $(c-a)^2(c-b)^2$ when $\epsilon \leq (c-a)^2(c-b)^2$. So f is discontinuous at c when $c \neq a, b$. Meanwhile, f is continuous at both a and b, just as with the function in section 15.3 in the main text. Specifically, since $f(a) = f(b) = 0$, for every $\epsilon > 0$ we can find δ such that both branches of the function f are within ϵ of 0 whenever x is within δ of either a or b. And so f is continuous at exactly those two points. ☐

Actually, one can use $(x-a)(x-b)$ in place of $(x-a)^2(x-b)^2$ for the same effect. This will be a parabola crossing the x-axis transversely at a and b, rather than kissing tangently at a and b as the function f did above.

15.5 Suppose that we had defined continuity of a function f using $\leq \epsilon$ and $\leq \delta$ in place of $< \epsilon$ and $< \delta$, respectively. Would it be an equivalent definition?

Yes, it would be equivalent to use $\leq \epsilon$ and $\leq \delta$ in place of $< \epsilon$ and $< \delta$ in the definition of continuity.

Theorem. *For every function $f : \mathbb{R} \to \mathbb{R}$, the following are equivalent:*

1. *f is continuous. That is, $\forall c\, \forall \epsilon > 0\, \exists \delta > 0\, \forall x\, (\,|x - c| < \delta \implies |f(x) - f(c)| < \epsilon\,)$.*
2. *f satisfies the proposed alternative: $\forall c\, \forall \epsilon > 0\, \exists \delta > 0\, \forall x\, (\,|x - c| \leq \delta \implies |f(x) - f(c)| \leq \epsilon\,)$.*

Proof. $(1 \to 2)$ Assume that f is continuous. To establish statement (2), fix c and $\epsilon > 0$. By continuity there is $\delta > 0$ for which $|x - c| < \delta$ implies $|f(x) - f(c)| < \epsilon$ for every x. Let $\delta' = \delta/2$. Now observe that if $|x - c| \leq \delta'$, then of course $|x - c| < \delta$, and consequently $|f(x) - f(c)| < \epsilon$, and hence also $|f(x) - f(c)| \leq \epsilon$. So statement (2) is fulfilled by using δ'.

$(2 \to 1)$ Suppose that f satisfies the property of statement (2). In order to show that f is continuous, fix any c and any $\epsilon > 0$. It follows that $\epsilon/2 > 0$ as well, and so by statement (2) there is a number $\delta > 0$ for which $|x - c| \leq \delta$ implies $|f(x) - f(c)| \leq \epsilon/2$ for every x. In particular, if $|x - c| < \delta$, then also $|x - c| \leq \delta$ and so $|f(x) - f(c)| \leq \epsilon/2$, and consequently $|f(x) - f(c)| < \epsilon$. So f is continuous. $\square$

15.7 Prove that every bounded nonempty set A of real numbers has a greatest lower bound.

Theorem. *Every bounded nonempty set of real numbers has a greatest lower bound.*

Proof. Suppose that $A \subseteq \mathbb{R}$ is a bounded nonempty set of real numbers. Let b be a lower bound of A, meaning that $b \leq a$ for every $a \in A$. Let

$$B = \{\, c \mid b \leq c \text{ and } c \text{ is a lower bound of } A \,\}.$$

This is a bounded nonempty set of real numbers, since $b \in B$ and every $c \in B$ is between b and any particular element of A. So B has a least upper bound $c = \sup(B)$. I claim that c is a lower bound of A, meaning that $c \leq a$ for every $a \in A$. If this fails, then there is some $a \in A$ with $a < c$. But since c is the least upper bound of B, there must be some $d \in B$ with $a < d < c$. But since $d \in B$, it is a lower bound of A, and so it cannot be that $a < d$ for some $a \in A$. So c must be a lower bound of A after all. In this case, it must be the greatest such lower bound of A, because c is the supremum of all such lower bounds, and so there can be none above c. $\square$

15.9 Does the least-upper-bound property hold for the rational numbers?

No, it does not.

Theorem. *The rational numbers do not enjoy the least-upper-bound property.*

Proof. Let $A = \{ q \in \mathbb{Q} \mid 0 < q \text{ and } q^2 \leq 2 \}$. This is the set of rational numbers in the interval $(0, \sqrt{2})$. The least upper bound in the real numbers would be $\sqrt{2}$, but this is not available in the rational numbers $\mathbb{Q}$. The upper bounds of A in $\mathbb{Q}$ are precisely the rational numbers above $\sqrt{2}$, but there is no least such number, since every nontrivial interval in $\mathbb{R}$ contains a rational number. So A has no least upper bound in $\mathbb{Q}$, and therefore the rational numbers do not satisfy the least-upper-bound property. $\square$

15.10 Prove that every triangle in the plane has a line that bisects both the area and the perimeter simultaneously. Is it necessarily unique?

Lemma. *For any triangle T in the plane and any angle θ, there is a line ℓ making angle θ with respect to the x-axis and bisecting the area of T, and furthermore, this line is unique.*

Proof. Consider a fixed line ℓ_0 at the desired angle but passing completely below the triangle. Let ℓ_t be the line obtained by vertically displacing ℓ_0 by distance t. There is some large enough distance k so that ℓ_k lies totally above the triangle. Let $f(t)$ be the fraction of the area of T below ℓ_t. So $f(0) = 0$ and $f(k) = 1$. Notice that the function f is continuous, since by insisting that displacements are very tiny, we can thereby ensure that the change in the area is as small as we like, and consequently, the change in f can also be guaranteed to be as small as we like. By the intermediate-value theorem, therefore, there is some value of t for which $f(t) = 1/2$. In other words, line ℓ_t exactly bisects the area of the triangle, as desired. This area-bisecting line is unique, since any larger value of t will have too much area below, and any smaller value of t will have too much area above. $\square$

Theorem. *Every triangle in the plane has a line that bisects both the area and the perimeter simultaneously.*

Proof. For each angle θ, let $\ell(\theta)$ be the unique line that makes angle θ with the x-axis and bisects the area of the triangle. For each angle θ, we view the line $\ell(\theta)$ as having a right side and a left side, as determined by one's right and left hands, if one should follow the angle θ out of the origin. For example, when the angle is 0, then we use the portion below $\ell(0)$ as the right side, and for $0 < \theta < 90°$, we continue to use the portion below $\ell(\theta)$ as the right side. When $90° \leq \theta < 180$, this is better described as the half plane to the right or above $\ell(\theta)$. At $\theta = 180°$, we are naturally using the upper half plane as the right side.

Let $p(\theta)$ be the proportion of the perimeter of the triangle that is on the right side of the line $\ell(p)$. Notice that this is a continuous function, since if some small change in the proportion of the perimeter is desired, one can ensure that by making only sufficiently small changes in the angle.

Finally, the key thing to notice is that, since $\ell(\theta)$ and $\ell(\theta+180°)$ are the same line but with their right and left sides swapped, it follows that $p(\theta + 180°) = 1 - p(\theta)$. In particular, if $p(\theta) < 1/2$, then $p(\theta+180°) > 1/2$, and so in any case, by the intermediate-value theorem, there must be some angle α with $p(\alpha) = 1/2$. Therefore, the line $\ell(\alpha)$ bisects both the area and the perimeter of the triangle. □

Finally, we note that the simultaneous-bisecting line need not be unique. For example, an equilateral triangle has three such lines, the angle bisectors of any of the corners.

15.11 Generalize the Heine-Borel theorem to arbitrary closed intervals $[a, b]$ in the real numbers. That is, prove that if $\mathcal{U}$ is a set of open intervals covering $[a, b]$, then there are finitely many intervals in $\mathcal{U}$ that cover $[a, b]$.

Theorem. *If $\mathcal{U}$ is a set of open intervals that cover a closed interval in the real numbers $[a, b]$, then there are finitely many intervals in $\mathcal{U}$ that cover $[a, b]$.*

Proof. Assume that $\mathcal{U}$ is a set of open intervals and that every $x \in [a, b]$ is an element of some $U \in \mathcal{U}$. Let B be the set of $x \in [a, b]$ for which the interval $[a, x]$ is covered by the union of finitely many elements of $\mathcal{U}$. So $x = a$ itself is in B, since $[a, a] = \{a\}$ and this is covered by some element of $\mathcal{U}$. So B is nonempty. Let d be the least upper bound of B. So $d \leq b$, since $B \subseteq [a, b]$. We know that $d \in U$ for some $U \in \mathcal{U}$. The set U has the form (r, s), with $r < d < s$. Since d is the least upper bound of B, there is some $c \in B$ with $r < c < d$. Since $c \in B$, we know that $[a, c]$ is covered by finitely many elements of $\mathcal{U}$, so $[a, c] \subseteq U_0 \cup \ldots \cup U_n$, with each $U_i \in \mathcal{U}$. But $[c, d] \subseteq (r, s) = U$, and so we have covered the interval $[a, d] = [a, c] \cup [c, d] \subseteq U_0 \cup \ldots \cup U_n \cup U$ by simply adding the set U. So we have proved $d \in B$. But furthermore, if $d < b$, then this finite cover will extend slightly beyond d up to s, since U is (r, s). Therefore, there would be elements of B larger than d, but this would contradict the fact that d is the least upper bound of B. And so it must be that $d = b$, and so the whole interval $[a, b]$ admits a finite cover from $\mathcal{U}$, as desired. □

We gave a direct independent proof, but one could alternatively prove the Heine-Borel theorem for arbitrary closed intervals $[a, b]$ as a consequence of the theorem for the unit interval $[0, 1]$, simply by scaling. That is, given an open cover of $[a, b]$, apply the affine transformation mapping this to $[0, 1]$ in order to produce an open cover of $[0, 1]$, get a finite subcover there, and then transfer those sets back to $[a, b]$. In this way, one will have found a finite subcover of the original closed interval.

> **15.13** Give an alternative proof of the Heine-Borel theorem using the principle of continuous induction.

Theorem. *Every open cover of a closed interval $[a, b]$ admits a finite subcover.*

Proof. Suppose that $\mathcal{U}$ is a set of open sets covering $[a, b]$. Let us prove by continuous induction that for every x in the interval, the interval $[a, x]$ is covered by finitely many elements of $\mathcal{U}$. This is true for $x = a$, since $[a, a]$ has only one point, which is covered by some element of $\mathcal{U}$. Suppose that $[a, x]$ is covered by $U_0 \cup \ldots \cup U_n$, with each $U_i \in \mathcal{U}$. Since these are open sets, the union must stick a bit beyond x, since if $x \in U_i$, then there is $\delta > 0$ with $[x, x + \delta] \subseteq U_i$. So we have also covered $[a, x + \delta]$ with finitely many sets from $\mathcal{U}$. Finally, suppose that x is a number such that for every $y \in [a, x)$ we can cover $[a, y]$ with finitely many sets from $\mathcal{U}$. Since $x \in U$ for some $U \in \mathcal{U}$, there is some $y < x$ with $[y, x] \subseteq U$, and so we can cover $[a, x]$ by augmenting the cover of $[a, y]$ with the set U, which will form a cover of $[a, x]$. Thus, by continuous induction, we conclude that every x in $[a, b]$ has the property that $[a, x]$ is covered by finitely elements of $\mathcal{U}$. Since this includes the case $x = b$, we have thereby proved that $[a, b]$ is covered by finitely many elements of $\mathcal{U}$, as desired. $\square$

> **15.15** In the text, we used the least-upper-bound principle to prove the continuous induction principle. Prove the converse. That is, assume the principle of continuous induction, and derive the least-upper-bound principle as a consequence.

Theorem. *Assume the continuous induction principle. Then the least-upper-bound principle holds.*

To clarify the dialectic here, we intend to prove the theorem *without assuming the least-upper-bound principle*. The point is that these two principles are equivalent as axioms and that in the logical development of the subject of real analysis, one can assume either one of them as a fundamental principle, since each one implies the other.

Proof. Assume the continuous induction principle, and suppose that $A \subseteq \mathbb{R}$ is a nonempty bounded set of real numbers. Assume toward contradiction that A has no least upper bound. Fix a particular element $a \in A$. Since A has no least upper bound, there must be other elements of A above a, and so a is not an upper bound of A. If a real number x is not an upper bound of A, then there is a larger element b in A above x, and so none of the numbers strictly between x and b are upper bounds of A. If a real number x has the property that every $y < x$ is not an upper bound of A, then x itself cannot be an upper bound of A, for then it would be the least upper bound, which we assumed didn't exist. So we have proved by continuous induction that every $x \geq a$ is not an upper bound of A. But this contradicts our assumption that indeed A had some upper bounds. So A must have a least upper bound after all. □

15.17 Prove the following continuous induction principle for closed intervals: Suppose that $a < b$ and that $B \subseteq [a, b]$ has the properties (1) $a \in B$; (2) whenever $r \in B$ and $r < b$, then there is $\delta > 0$ with $[r, r + \delta) \subseteq B$; and (3) whenever $[0, r) \subseteq B$, then $r \in B$. Then $B = [a, b]$ is the entire closed interval.

This was the form of continuous induction that we already used in exercises above, but let us now provide a proof of it. We assume the least-upper-bound principle.

Theorem. *If $B \subseteq [a, b]$ is a set of real numbers in the interval $[a, b]$ such that*

1. *$a \in B$;*
2. *whenever $x \in B$, then either $x = b$ or there is some $\delta > 0$ with $[x, x + \delta) \subseteq B$; and*
3. *whenever x is a nonnegative real number and $[0, x) \subseteq B$, then $x \in B$,*

then B is the whole interval $[a, b]$.

Proof. We prove the principle using the least-upper-bound property of the real numbers. Let C be the set of $x \in [a, b]$ for which $[a, x] \subseteq B$. This includes the case $x = a$, since $[a, a] = \{a\}$, which is contained in B since $a \in B$ by statement (1). So $a \in C$, and so in particular, C is nonempty. Let d be the least upper bound of C. This implies that $[a, d) \subseteq B$, because if $y \in [a, d)$, then there is some $r \in C$ with $y < r < d$, and consequently $[a, r] \subseteq B$, and hence $y \in B$. So by hypothesis (3), we see that $d \in B$. If $d \neq b$, then by hypothesis (2), it follows that $[d, d + \delta) \subseteq B$ for some $\delta > 0$. But this means that $[a, d + \delta/2] \subseteq B$, which places $d + \delta/2$ into C, contrary to our assumption that d was the least upper bound of C. So it must be that $d = b$, and consequently $[a, b] \subseteq B$, and hence $B = [a, b]$, as desired. □